John F. Hoffecker
Scott A. Elias

Human Ecology
of Beringia

COLUMBIA UNIVERSITY PRESS NEW YORK

Columbia University Press
Publishers Since 1893
New York   Chichester, West Sussex
Copyright © 2007 Columbia University Press
All rights reserved
Library of Congress Cataloging-in-Publication Data
Hoffecker, John F.
Human ecology of Beringia / John F. Hoffecker Scott A. Elias.
p. cm.
Includes bibliographical references and index.
ISBN-13: 978-0-231-13060-8 (cloth : alk. paper)
1. Human ecology—Bering Land Bridge. I. Elias, Scott A. II. Title.
GF891.H64 2007
304.20957′709012—dc22
200638539

Columbia University Press books are printed on permanent and durable acid-free paper.
Printed in the United States of America
c 10 9 8 7 6 5 4 3 2 1

Dedicated to the memory of

David M. Hopkins
(1921–2001)

First citizen of Beringia

Contents

Preface
Lost Continent

Public fascination with Atlantis and other mythical "lost continents" has a long history. Ironically, interest in the real thing—Beringia—has been left largely to scientists. The lowered sea levels of the Ice Age exposed coastal margins and several large land areas around the world, but their most significant consequence for humanity was the immense plain that emerged between Northeast Asia and Alaska. Across that plain, most probably, walked the first people of the New World.

Beringia has its own story to tell, however, quite apart from its historic role in the peopling of the Western Hemisphere. Originally defined strictly in terms of the now submerged land bridge between Asia and America, Beringia has grown in the minds of those who study it to encompass a much larger area. As defined here, it stretches more than 4,000 kilometers from the Verkhoyansk Mountains in the west to the Mackenzie River in the east. Beringia was truly continental in size, and most of it lay above latitude 60°N. It was a land of mountain glaciers and frozen lowlands, isolated to a significant degree from other parts of the earth.

We have written a synthesis of the human side of Beringia. It is a comparatively brief slice of prehistory, because humans seem to have come to Beringia during its final days—only a few millennia before rising seas flooded the Bering Strait for the last time. We have addressed the question of why people came to Beringia so late in prehistory: Why did people not come earlier or, for that matter, later? And we have described the twilight world that the early Beringians inhabited—after plants and animals of the shrub tundra spread across the landscape, but before the sea reclaimed the central plain. Much of their archaeological record is uniquely Beringian and reflects this isolated world. We have also recounted the shattering impact of the Younger Dryas period, which saw the end of Beringia and major changes in the archaeological record. Some of these changes have been a source of confusion among archaeologists for many years. Finally, we have addressed the thorny issue of how the prehistory and human ecology of Beringia are related to the peopling of the New World.

Both of us are indebted to a number of people in many fields who have contributed in one way or another to the making of this book. Some of them are people with whom one or both of us have collaborated, while others are colleagues who have generously shared information and ideas. Specifically, we thank Robert E. Ackerman, Thomas A. Ager, G. F. Baryshnikov, Nancy H. Bigelow, Peter M. Bowers, Julie Brigham-Grette, Ariane Burke, Jacques Cinq-Mars, Barbara Crass, E. James Dixon, Don E. Dumond, Norman A. Easton, Mary E. Edwards, Ted Goebel, R. Dale Guthrie, Thomas D. Hamilton, Gary Haynes, Charles E. Holmes, Michael L. Kunz, William S. Laughlin, Craig M. Lee, Ralph A. Lively, William F. Manley, Owen K. Mason, Vladimir Pitul'ko, William Roger Powers, Richard D. Reger, G. Richard Scott, Susan K. Short, Tim A. Smith, Norman W. Ten Brink, Robert M. Thorson, Richard VanderHoek, N. K. Vereshchagin, Christopher F. Waythomas, Frederick Hadleigh West, William B. Workman, and David R. Yesner.

Several colleagues have contributed directly to the writing of this book, by either providing unpublished information or reviewing chapters in draft form, or both. We thank the following: Charles E. Holmes for providing information on the most recent work at Swan Point and for reviewing chapter 4; David R. Yesner for information on the fauna from the Tanana Basin sites; Gennady Baryshnikov for information on Berelekh and the photograph from this important site; Norm Easton for new information on the Little John site in the southern Yukon;

S. B. Slobodin for information on sites in the Upper Kolyma Basin; Owen Mason for reviewing chapter 6; and Ted Goebel for reviewing all chapters and for information on Ushki and several other sites. Finally, we thank Craig M. Lee for preparing box 4.2 on microblades. Figures 1.1, 3.8, 3.9, 4.2, 4.9, 4.11, 5.1, 5.4, and 6.1–6.5 were prepared by Ian Torao Hoffecker.

We have dedicated this book to the memory of David M. Hopkins, who was an inspiration to all students of Beringia, including the authors.

John F. Hoffecker
Scott A. Elias

Human Ecology of Beringia

An Introduction to Beringia

During the summer of 1985, the research vessel *Discoverer* was cruising in waters of the Chukchi Sea roughly 500 kilometers north of the Bering Strait. The *Discoverer* was a 92-meter-long diesel-powered ship owned by the National Oceanographic and Atmospheric Administration and used for various research projects.[1] In 1985, it was collecting sediment cores from the submerged continental shelf between Chukotka and northern Alaska. The project was under the direction of Lawrence Phillips of the U.S. Geological Survey (Phillips and Brouwers 1990).

Phillips drilled more than twenty cores into the shallow floor of the Chukchi Sea and brought them back to the U.S. Geological Survey center at Menlo Park in California. Beneath a layer of recent marine sediment, the cores were found to contain peat deposits of the Bering Land Bridge that once joined Northeast Asia and Alaska. Samples of the peat were shipped to researchers at the Institute of Arctic and Alpine Research at the University of Colorado for analysis of fossil beetle remains, plant fragments, and pollen (Elias, Short, and Phillips 1992). The fossil biota suggested a grassy shrub tundra environment with dwarf birch and willow under climates that were somewhat warmer than those of the present day. The cores retrieved by the *Discoverer* presented a snapshot of the northern edge of the Bering Land Bridge during its final centuries—shortly before rising sea levels flooded the lowlands between Chukotka and Alaska, separating North America from Eurasia.

The most surprising information produced by the cores was the age of the peat. Calibrated radiocarbon dates indicated that the tundra plants and animals were only 12,900 years old—several thousand years younger than the widely accepted dates for the Bering Land Bridge (Elias, Short, and Phillips 1992:373–374; Elias et al. 1996).[2] The dates from the cores provided the basis for a new land bridge chronology, which was consistent with a recently published high-resolution sea-level chronology from the Caribbean (Fairbanks 1989). They revealed that the land bridge was still present as climate changed at the end of the last glacial period, transforming the biota of the region from an arid steppe-tundra populated by mammoth, bison, and other large grazing mammals to a wet shrub tundra inhabited by modern fauna (Guthrie 1990).

The new land bridge chronology also had major implications for archaeology because it confirmed that the earliest known sites in Alaska were occupied when it was still possible to walk from Asia to North America (Hoffecker, Powers, and Goebel 1993:46–47; West 1996a:551).[3] Humans moved onto the Bering Land Bridge and adjoining parts of Northeast Asia and Alaska during a twilight period at the end of the Ice Age that was unlike the periods that preceded and followed it. The character of the shrub tundra environment in this part of the world— and the unique archaeological record that comes with it—is a central theme of this book.

Defining Beringia

The existence of a northern land connection between Asia and America was first proposed in AD 1590 by the Spanish missionary Fray José de Acosta, who suggested that it provided a bridge for the peopling of the New World (Wilmsen 1965). Even for the sixteenth century, which was a remarkable period of geographic and scientific exploration (Boorstin 1983), this was an extraordinary insight. It was more than a century later

(AD 1728) that Vitus Bering sailed into the narrow strait between the two continents that bears his name.

During the nineteenth century, as knowledge of the geography of the Bering Strait region increased, there was further speculation about a connection between Siberia and Alaska (Hopkins 1967a:1–3). Although some naturalists, including Charles Darwin (1859:365–382), concluded that the entire circumpolar region had been covered by ice during the last glacial epoch, G. M. Dawson suggested that much of Alaska had been unglaciated and joined to Northeast Asia by a "wide terrestrial plain" (Dawson 1894:143–144). This speculation was inspired by the shallow depths of the Bering and Chukchi Seas and bolstered by discoveries of mammoth remains on the Aleutians and Pribilof Islands (Dall and Harris 1892; Stanley-Brown 1892).

The distribution of biota on opposing sides of the Bering Strait became an important part of the debate during these years. The nearly identical flora and fauna of the Eurasian and American Arctic zones offered the strongest argument for a recent land connection across the shallow strait (e.g., Wallace 1876; Heilprin 1887). Biogeography ultimately became a primary basis for the concept of Beringia.

In the early years of the twentieth century, Adolph Knopf pieced together a general history of the Bering Land Bridge from various sources of geologic data (Knopf 1910). At the time, the tectonic uplift and subsidence of land masses seemed the most likely mechanism for the past emergence of the land bridge (Hopkins 1967a:2). By the 1930s, however, geologists had recognized the link between glaciation and ocean volume (Daly 1934) and concluded that the Bering Land Bridge was a product of expanding ice sheets and falling sea levels (Smith 1937).

The term *Beringia* was first proposed by the Swedish botanist Eric Hultén in a 1937 book titled *Outline of the History of Arctic and Boreal Biota During the Quaternary Period*. Hultén (1937:34) applied the term to the now-submerged areas of the continental shelf between Northeast Asia and Alaska—that is, the Bering Land Bridge (e.g., Hopkins 1967a:3; West 1981:1; Matthews 1982:128). He embraced the notion of glacier growth and consequent sea-level reduction as the mechanism that created Beringia, and he was especially interested in the effect of both on the distribution of plants and animals (Hultén 1937, 1968) (figure 1.1).

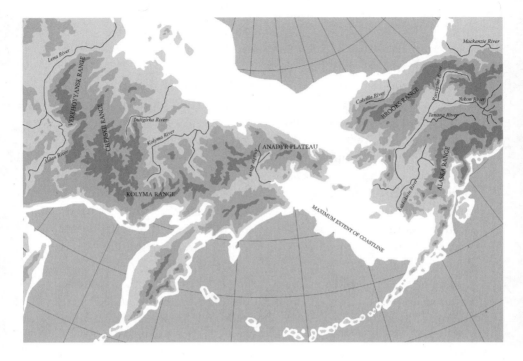

Figure 1.1 Beringia, illustrating boundaries discussed in the text

The late David M. Hopkins, who devoted a long and productive career to the study of Beringia, broadened its geographic definition to include western Alaska and portions of Northeast Asia as far west as the Kolyma River (Hopkins 1967b:vii). Hopkins and others later expanded the eastern boundary to the lower Mackenzie River in the Northwest Territory of Canada (e.g., Hopkins, Smith, and Matthews 1981:218–219, fig. 1; Matthews 1982:128). This aligned the eastern edge of Beringia with the margin of the Laurentide ice sheet at its maximum extent where it was an impenetrable barrier to plants and terrestrial animals. Although the Kolyma River represents a less formidable ecological boundary, Russian paleobotanist Boris Yurtsev noted that it marks the western limit of a continental Beringian flora—the "Yukon-Kolyma species" (Yurtsev 1984:134–135).[4]

Several archaeologists proposed extending the western boundary much farther into Northeast Asia. Frederick Hadleigh West favored a definition of Beringia that encompassed a significant portion of the Lena Basin (West 1981:2–3), corresponding broadly to Yurtsev's concept of Megaberingia that was based on the distribution of relict flora (Yurtsev 1984:150–151, fig. 11).[5] This provided a more workable definition of Beringia from an archaeologist's perspective (e.g., Morlan 1987:268) because there were (and remain) few known sites of Pleistocene age between the Lena Basin and Alaska (Goebel and Slobodin 1999).

In the early 1990s, Hoffecker, Powers, and Goebel (1993:46) proposed the Verkhoyansk Mountains as a more suitable western boundary for Beringia (see also Goebel and Slobodin 1999:105–106). The Verkhoyansk Range, which lies along the eastern margin of the lower Lena Basin, has some climatic significance (e.g., boundary of 60 days or less of temperatures above 0°C [*New Oxford Atlas* 1978:98]) and delineates the western limit of some characteristic Beringian plant taxa (e.g., *Polygonum tripterocarpum* [Yurtsev 1984:136, fig. 5]). It may or may not have been a major barrier to human settlement,[6] but it does appear that much of the land between the Verkhoyansk Mountains and the Laurentide ice sheet was colonized as a unit during the Lateglacial (see chapter 4).

Sea-Level History

Since the 1930s, it has been apparent that Beringia was formed and dissolved by changes in global sea level—and indirectly by the changes in glacier volume that controlled the latter. The chronology of sea-level history is therefore fundamental to the history of Beringia, and the recent revisions of that chronology have greatly affected the study of human ecology in Beringia.

Global sea-level history, however, is only part of the story. The other part is the geomorphology of the shallow Bering-Chukchi Platform that became the Bering Land Bridge during episodes of lowered sea level. The platform is a "monotonously flat" submerged plain that extends 1,600 kilometers from the Arctic Ocean to the eastern Aleutians (Hopkins 1959:1519).[7] A thin veneer of recent marine sediment fills minor troughs and contributes to the largely featureless topography (Gershanovich 1967:39–42). Isolated high points are represented by St. Lawrence Island, Wrangel Island, and a few other small islands in the Bering and Chukchi Seas.

Along the southern edge of the plain is a steep submarine escarpment that drops roughly 1,500 meters to the floor of the western Bering Sea, while the slope on the northern margin descends more gently into the Arctic Ocean basin floor (Hopkins 1959:1519). Most of the continental shelf lies less than 100 meters below the present-day sea level. Depths near the southeast tip of the Chukotka Peninsula are especially shallow (approximately 30 meters), and a land connection with St. Lawrence Island survived into the early Holocene epoch (Hopkins 1967c:464–465). When sea level fell roughly 50 meters or more below that of the present day, it exposed a land bridge between Chukotka and Alaska.

Before the end of the Miocene epoch, Northeast Asia and Alaska were parts of one continental mass that separated the Pacific and Arctic Oceans. Between 5.5 and 4.8 million years ago (mya), depression of the Bering-Chukchi Platform opened the strait for the first time, allowing a major exchange of marine molluscan fauna between the Pacific and Arctic-Atlantic Basins (Hopkins 1967c:452–460; Marincovich and Gladenkov 2001). During the early and middle Pleistocene (1.8–0.13 mya), the marine isotope record indicates that—with the Bering-Chukchi Platform more or less in its present position—the Bering Land Bridge was repeatedly exposed and inundated in response to oscillating climates and glacier volumes (Shackleton and Opdyke 1973).

During the Last Interglacial climatic optimum (128,000–116,000 years ago), both temperatures and ocean levels rose above those of the present day (Marine Isotope Stage 5e [MIS 5e] age equivalent; box 1.1). The land bridge was completely flooded, and beach ridges were formed as much as 7 meters above modern sea level along the north Alaskan coast (Hopkins 1982:12–13; Brigham-Grette and Hopkins 1995; Brigham-Grette et al. 2001). Climates remained warm throughout most of the following 40,000 years (115,000–75,000 years ago; figure 1.2), and the land bridge was probably exposed only briefly during this interval (MIS 5d–5a). A significant drop in sea level occurred roughly 70,000–60,000 years ago, restoring the land bridge (MIS 4 age equivalent) (Hopkins 1982:13).[8]

The lengthy interstadial period that is equated with MIS 3 (or the Middle Wisconsin period in North America) has been difficult to reconstruct in terms of land bridge history (Hopkins 1982:13). Between ca. 60,000 and 28,000 years ago, temperatures in the Northern

Box 1.1
Framework for the Pleistocene: Marine Isotope Stages

In a departure from the traditions of geology, the stratigraphic framework for the Pleistocene (i.e., the Ice Age) is based on a sequence of marine isotope stages that reflect past climate change.

An isotope is a variety of an element that has the same number of protons but a different number of neutrons. While the various isotopes of an element have the same chemical attributes, they often display different physical attributes (e.g., carbon-12, which is stable, and carbon-14, which is radioactive). The isotope of an element is specified by the sum of the number of protons and neutrons. Thus, ^{16}O has eight protons and eight neutrons, and ^{18}O has eight protons and ten neutrons.

Oxygen is the most abundant element in the Earth's crust. It accounts for 89 percent of the mass of the ocean, and it is the second most abundant element in the Earth's atmosphere. Oxygen has two stable isotopes: ^{16}O and ^{18}O. ^{16}O is by far the most abundant isotope, accounting for 99.76 percent of the oxygen on the planet. The heavier isotope ^{18}O accounts for only 0.24 percent of the Earth's oxygen.

Due to the difference in mass, there is a fractionation of water molecules containing the two oxygen isotopes during the processes of evaporation, condensation, and precipitation. When water is evaporated from the ocean into the atmosphere, it has more ^{16}O than the ocean has because ^{16}O bonds are slightly weaker than ^{18}O bonds. Thus the isotopic composition of water vapor in the atmosphere is not the same as in the ocean. There is fractionation in the opposite direction when precipitation (rain or snow) causes the water to condense. The heavier isotope water condenses more easily than the lighter isotope variety.

The buildup of ice into continental ice sheets causes a net loss of ^{16}O from the oceans. Seawater becomes isotopically "heavy" during glacials (i.e., enriched in ^{18}O). During interglacials, the melting of ice sheets releases ^{16}O waters back into the oceans, making them isotopically "lighter." The history of these oxygen-isotope variations is preserved in the fossil shells of marine organisms, such as foraminifera and marine mollusks. The isotopic ratio between ^{18}O and ^{16}O is expressed in terms of the difference between this ratio in a given sample and the ratio found in modern seawater. This ratio is termed $\delta^{18}O$. In 1946, Harold Urey recognized that the oxygen-isotope ratios in calcium carbonate in marine shells depend on the temperature. Experiments suggested that a 1°C decline in temperature brings about a 0.2 per mil increase in $\delta^{18}O$ in foram and mollusk shells. Thus the stage was set for the investigation of ancient sea surface temperatures through the

(Box 1.1 continued)

analysis of δ¹⁸O values preserved in marine microfossils made of calcium carbonate.

An important breakthrough took place in 1973, when Shackleton and Opdyke published a long record of oxygen isotopes from marine microfossils out of a deep sea core taken from the equatorial Pacific. This δ¹⁸O record showed large-scale changes that mirrored the timing of glacial/interglacial cycles as predicted by the orbital-forcing model of Milankovitch. Shackleton and Opdyke's study also showed that isotope ratios had changed throughout the world's oceans in response to glacial/interglacial cycles, not simply in the seas that were in close proximity to ice sheets. Shackleton and Opdyke proposed a numbering scheme for major changes in marine isotope ratios, called marine isotope stages (MIS). They identified the current interglacial (the Holocene) as MIS 1 and worked back through time into the Pleistocene, assigning even numbers to glacial stages (the Last Glacial Maximum [LGM] is equivalent to MIS 2) and odd MIS numbers to interglacial and major interstadial events. MIS 3 represents the interstadial warming during the last glaciation, and MIS 5 represents the Last Interglacial. Figure 1.2 shows a δ¹⁸O record for the North Atlantic that extends back throughout the Quaternary Period (the last 2.6 million years). There have been approximately 100 MIS (or 50 glacial/interglacial oscillations) in the Quaternary.

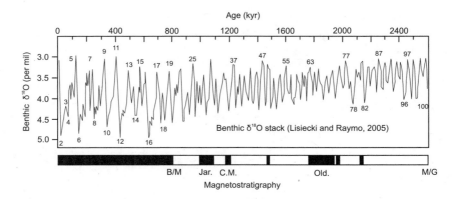

Figure 1.2 δ¹⁸O record for the North Atlantic for the past 2.6 million years (after Lisiecki and Raymo 2005)

Hemisphere were generally lower than those of the present day and constantly oscillated from cool to warm (Dansgaard et al. 1993). Sea level was also lower as a consequence but may have fallen no more than 70 meters during the coldest oscillation and generally remained less than 50 meters below that of today (Anderson and Lozhkin 2001: 95–96).

The Bering Land Bridge appears to have been absent during much or most of the interstadial period correlated with MIS 3. This conclusion has acquired some potential significance for the history of human settlement in Beringia because of the recent discovery of an archaeological site in the lower Yana River Basin dating to roughly 30 cal ka (30,000 calibrated radiocarbon years ago) (Pitul'ko et al. 2004) and indicating occupation—at least on a seasonal basis—of areas east of the Verkhoyansk Mountains at the end of the interstadial. Sites of comparable age in northern Europe reveal a similar pattern of high-latitude settlement (Pavlov, Svendsen, and Indrelid 2001). The final phase of the MIS 3 interstadial was relatively mild (conditions in the Yana Lowland may have been similar to those of today [Anderson and Lozhkin 2001:98–99]), and it appears likely that a land connection to Alaska was not present at the time.

During the last glacial (or MIS 2 age equivalent), massive ice-sheet growth occurred in the Northern Hemisphere, and global sea level fell to a minimum of 120 meters below that of the present day (Hopkins 1982:14; Fairbanks 1989). The entire Bering-Chukchi Platform was exposed, creating a land bridge more than 1,000 kilometers wide during the Last Glacial Maximum (23–19 cal ka). In contrast to the preceding interstadial period, humans probably did not occupy latitudes above 55°N at this time (Goebel 1999:218),[9] despite the fact that Beringian environments supported a rich assemblage of large mammals (Matthews 1982; Guthrie 1982, 1990). Identifying the variable(s) that apparently excluded settlement during the Last Glacial Maximum is a central issue in the human ecology of Beringia (figure 1.3).

Although previous research on land-bridge history suggested that rising ocean levels had flooded the strait as early as 17–16 cal ka (McManus and Creager 1984), the improved estimates of sea-level chronology and dated peat deposits from the Chukchi Sea indicate a land connection as late as 12.9 cal ka (Fairbanks 1989; Elias et al. 1996). Ironically, final inundation of the land bridge seems to coincide with a brief cold oscillation that occurred at the end of the Pleistocene ca. 12.8–11.3 cal ka (Younger Dryas) and which significantly influenced human ecology in eastern Beringia (see chapter 6). As temperatures peaked at 2–3°C warmer than today during the early postglacial epoch, rising sea levels deposited high beach ridges along the shores of Chukotka and Alaska (e.g., Jordan 2001:513–518).

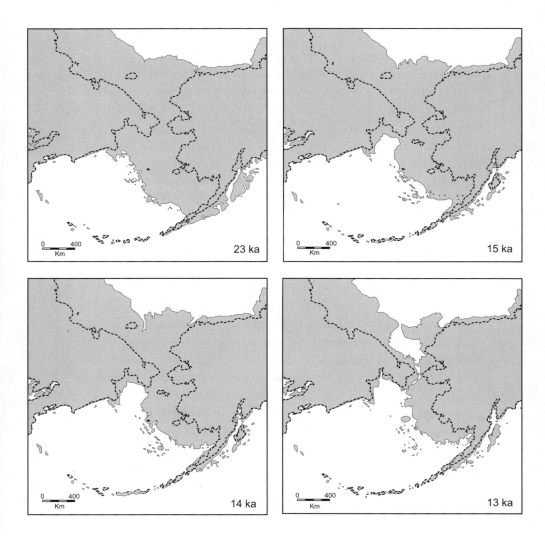

Figure 1.3 Sea-level history of Beringia since the Last Glacial Maximum (based on Manley 2002)

Glacial Chronology and the "Ice-Free Corridor"

The chronology of glacial advances and retreats is fundamental to the history of Beringia not only because of their effect on sea level and the land bridge but also because of their role in isolating Beringia from North America (e.g., Maddren 1905). At their maximum extent, the coalescing Laurentide and Cordilleran ice sheets completely blocked movement of terrestrial biota between eastern Beringia and mid-latitude North America. The retreat of these glaciers after 17 cal ka eventually opened up coastal and interior areas for human settlement.

The discovery of an early human occupation in the lower Yana River area has stimulated new interest in glacier advances before the Last Glacial Maximum (LGM). Would ice sheets have precluded movement from Beringia to unglaciated parts of North America at this time? The dating of glacial deposits in the Richardson and Mackenzie Mountains indicates that advancing Laurentide ice blocked access to southwestern Canada at the northern end of what would later become an ice-free corridor as early as ca. 30 cal ka (Catto 1996; Jackson and Duk-Rodkin 1996; Mandryk et al. 2001:302–304). At the time that people occupied Yana RHS (Pitul'ko et al. 2004), both water and ice seem to have been barriers to North America.

The timing of glacial retreat and the opening of coastal and interior routes to North America at the end of the LGM are of special importance to archaeologists because they were critical factors in the settlement of the New World. Research during the past fifteen years in Cook Inlet, Alexander Archipelago, Queen Charlotte Islands, and elsewhere suggests that deglaciation of the south Beringian and Northwest coasts took place earlier than previously thought—16 cal ka or slightly before (Mann and Peteet 1994; Hansen and Engstrom 1996; Dixon 2001:291). At the same time, dates on retreating glaciers in the Mackenzie Mountains currently indicate that an ice-free corridor did not open up until 15 cal ka at the earliest (Jackson and Duk-Rodkin 1996; Mandryk et al. 2001:303–304; Dyke 2004:411) (figure 1.4).

Most North American archaeologists accept a reported age of 15 cal ka for the occupation at Monte Verde in southern Chile (Dillehay 1989; Meltzer 1997). This holds significance for both the timing and routes of human migrations from Beringia to the New World. The dates on the ice-free corridor suggest that the interior route may not have been available to the people who occupied Monte Verde. Barring a much earlier migration before the LGM—for which there is little compelling evidence in either North or South America—the glacial chronology may indicate that the coastal route is more likely (Dixon 1999; Mandryk et al. 2001). The relevance of the Beringian data to the problem of New World origins is discussed in the final chapter of this book.

In recent years, Grosswald (1988) and Hughes and Hughes (1994) suggested that a massive ice sheet covered interior Beringia. The suggestion ran counter to a century of research and writing on Beringian glacial history (e.g., Hopkins 1967a:2–3) and has been emphatically rejected by other geologists on the grounds that moraines and other

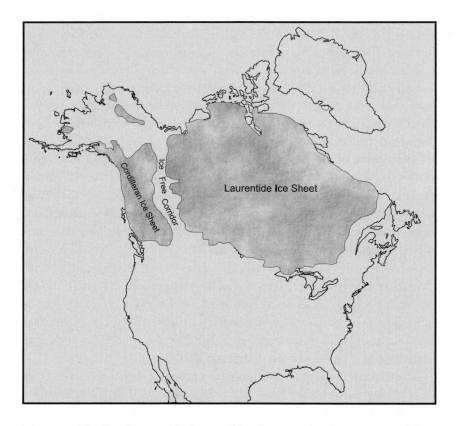

Figure 1.4 The "ice-free corridor" created by the retreating Laurentide and Cordilleran glaciers

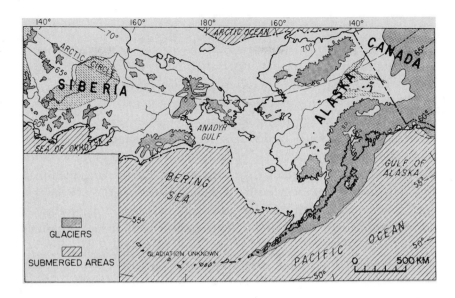

Figure 1.5 Glaciation in Beringia during the Last Glacial Maximum (fig. 3 from David M. Hopkins, *The Bering Land Bridge,* copyright © 1967 by the Board of Trustees of the Leland Stanford Jr. University, renewed 1995)

physical traces of glaciation are scarce in lowland areas outside the southern coast (Brigham-Grette 2001:18–19; Heiser and Roush 2001:403). During the LGM, glacier volume was low due to the extreme aridity of interior regions (Hopkins 1982:15). Beyond the southeastern coastal area—where moist air from the North Pacific fed the northern portions of the Cordilleran ice sheet—glaciers were almost entirely restricted to the Verkhoyansk Mountains, Cherskii Range, Koryak Upland, Brooks Range, and several other isolated uplands (e.g., Kind 1974:122–137; Hamilton and Porter 1975; Glushkova 2001; Heiser and Roush 2001) (figure 1.5).

The Environmental Debate

As in the case of sea-level history and glacial chronology, the discovery of an early human presence in the lower Yana River has encouraged archaeologists to take a fresh look at pre-LGM environments in Beringia. As already noted, climate conditions at the close of the MIS 3 interstadial were relatively mild, and pollen from lake sediments in the lower Kolyma

Basin dating to ca. 30 cal ka indicate a larch-birch forest with open boggy areas (Anderson and Lozhkin 2001:99). Tundra environments seem to have been more widespread in eastern Beringia at the time.

The principal debate over Beringian environments has been focused on the LGM period. Palynologists reconstructed herbaceous tundra and polar desert across much of Beringia (e.g., Colinvaux 1967; Ritchie and Cwynar 1982), but in a classic paper, paleontologist R. Dale Guthrie disputed this view on the basis of the remains of grazing mammals found in late Pleistocene deposits of central Alaska. Guthrie (1968:357–361) argued that the nutritional requirements of the mammals indicated a more productive steppe-tundra. Radiocarbon dates on large mammal bones and teeth confirm the presence of large numbers of grazers—especially bison, horse, and mammoth—in many parts of Beringia during the LGM (Matthews 1982:139–143; Guthrie 1990:239–245). Aridity and constant loess deposition probably were important factors in maintaining productive steppe-tundra vegetation at high latitudes under very cold conditions—habitat that has no modern analog (Guthrie 2001; Walker et al. 2001).

More recently, discussion has shifted to the question of regional variability. Based on dated pollen, plant macrofossils, and beetles from Chukchi and Bering Sea cores, Elias et al. (1996) postulated a mesic shrub tundra belt on the land bridge during the LGM. Guthrie (2001:560–565) thought that a central tundra zone might explain the absence of woolly rhinoceros in eastern Beringia—and a corresponding lack of several North American taxa in western Beringia (e.g., short-faced bear). But Ager (2003) argues that pollen samples from other localities on the former land bridge (e.g., St. Michael Island) do not indicate mesic tundra during the LGM. Yurtsev has consistently emphasized the variability of Beringian environments (e.g., Yurtsev 1982, 2001).

The scarcity of trees and woody shrubs during the LGM may have been a critical variable in human settlement (Guthrie 1990:273–277; Hoffecker, Powers, and Goebel 1993:48). Lack of adequate fuel—rather than food sources—is the most likely explanation of why humans occupied Beringia before and after the LGM, but not during the period 30–15 cal ka. Although the absence of LGM sites in Beringia may simply reflect limited sampling of regions that remain relatively unpopulated and undeveloped, it is consistent with a general pattern of high-latitude abandonment across Eurasia during this interval of extreme cold and aridity (Hoffecker 2005a:94–95).

Less contentious than the LGM environment debate—and of great importance to the problem of human settlement—is the shift to shrub tundra at the end of the LGM (i.e., beginning of the Lateglacial). Hopkins (1982:10–11) termed this period the Birch Zone because pollen samples from localities across Beringia invariably reveal a sudden increase in *Betula* pollen at 17–15 cal ka (e.g., Ager 1983; Bigelow and Powers 2001; Pisaric et al. 2001:238–241). It reflects a transition to warmer and wetter climates—correlated with a measurable rise in lake levels—and tundra vegetation with dwarf birch and willow (Bigelow and Edwards 2001; Mann et al. 2001).[10] Major changes in the fauna took place at this time—including the extinction of mammoth and horse—that were almost certainly tied to the changes in climate and vegetation (Guthrie 1990). The first permanent human settlement of Beringia also is tied to the shrub tundra transition (figure 1.6).

The warming trend was interrupted by a brief cold oscillation (Younger Dryas) that began at 12.8 cal ka. The effects of this cooling episode on Beringian environments are not always obvious in

Figure 1.6 Modern shrub tundra vegetation similar to the vegetation that covered much of Beringia during the Lateglacial (photograph by SAE)

the geologic and paleobiotic record and have been subject to debate. Lozhkin et al. (2001) found no evidence for colder conditions on Wrangel Island, and Szeicz and MacDonald (2001) did not see indications of lower temperatures in the Mackenzie Mountains during Younger Dryas time. In contrast, Elias (2001a) reported a strong Younger Dryas signal in dated samples of fossil beetles from northeastern Beringia, and Bigelow and Edwards (2001) found evidence for increased herb tundra taxa in pollen assemblages from south-central Alaska during 12.8–11.4 cal ka. Perhaps the most dramatic influence of the Younger Dryas may be found in the significant increase of steppe bison remains in Beringian archaeological sites—suggesting expanded steppic habitat.

Humans in Beringia

The study of Beringian archaeology has been constrained by several factors. The most important of these is the remoteness of the region and the lack of development—especially in western Beringia— during recent times. The sorts of construction activities that often yield archaeological discoveries (i.e., highways, reservoirs, and other industrial projects) are comparatively rare, and most of the region is relatively inaccessible (Hoffecker, Powers, and Goebel 1993:48). To date, only a handful of archaeological sites can be firmly dated to the earliest phase of permanent settlement (i.e., Lateglacial interstadial), while a slightly greater number of sites is known from the Younger Dryas interval. Large portions of Beringia are currently under water, and no sites have been discovered yet on the submerged coastal margins or central plain (Dixon 2001:282–290). Many of the known sites contain remains buried in shallow sediments (e.g., Mesa [Kunz, Bever, and Adkins 2003]) sometimes with little material suitable for radiocarbon dating (often reflecting a scarcity of wood), such as Bol'shoi El'gakhchan I (Kir'yak 1996). Many sites have been severely disturbed by frost action (e.g., Engigstciak [Mackay, Mathews, and MacNeish 1961:29–33]). Faunal remains are rare—frequently due to poor preservation conditions.

The human settlement of Beringia is inextricably linked to the settlement of Northeast Asia. Until humans developed the ability to cope with the extreme winter temperatures, comparatively low biological productivity, and other challenging features of Northeast Asian environments, they were unable to colonize the land east of the Verkhoyansk Mountains (Hoffecker and Elias 2003:36–37). Although evidence has been reported for a human presence at Diring Yuriakh on the Lena River at latitude 61°N before 260,000 years ago (Waters, Forman, and Pierson 1997), it remains controversial (e.g., West 1996a:542).[11] The occupation of the Altai region of southwest Siberia during ca. 130,000–50,000 years ago—presumably by the Neandertals—appears to be well documented (Goebel 1999:212–213). Despite morphological and ecological adaptations to northern environments, however, the cold tolerance of the Neandertals seems to have been more limited than that of modern humans (Hublin 1998; Hoffecker 2005a:55–59).[12]

Only modern humans (*Homo sapiens*) fully colonized Northeast Asia—during periods colder than today—and invaded arctic environments (Klein 1999). Paradoxically, the earliest modern humans to settle latitudes above 45°N were derived from the tropical zone and lacked the morphological adaptations to low temperature that were found among the Neandertals (Trinkaus 1981). They nevertheless occupied Northeast Asia as far as 55°N during 45–35 cal ka (Goebel 1999:213–216).

The critical variable probably was the ability of modern humans to design novel and sometimes complex technology—not only sewn clothing and artificial shelters, but also innovative food-getting implements and devices—to adapt to cold climates and scarce resources (Hoffecker 2005b). During this relatively mild interval toward the end of the MIS 3 interstadial, modern humans expanded into the European Arctic—at least on a seasonal basis (Pavlov, Svendsen, and Indrelid 2001). The discovery of an occupation dating to about 30 cal ka on the Lower Yana River reveals that they also intruded above the Arctic Circle in Northeast Asia (Pitul'ko et al. 2004) and had—according to the geographic definition followed here—moved into Beringia. The Bering Strait apparently was flooded at this time and, after 30 cal ka, the ice-free corridor was closed, blocking access to North America (Mandryk et al. 2001).

Despite dispersal across much of northern Eurasia during the later phases of the MIS 3 interstadial, modern humans may have abandoned portions of northern Europe and Northeast Asia during the coldest phase of the LGM (ca. 23–20 cal ka) (e.g., Soffer 1985:173–176; Goebel 1999:218). The reasons for temporary retreat from these areas remain obscure but might include the limitations of their technology, lack of adequate fuel sources, and retention of warm-climate body proportions (Hoffecker and Elias 2003:37–38; Hoffecker 2005b).[13] In any case, modern humans began to reoccupy Northeast Asia as climates ameliorated after 20 cal ka, and humans were present in the Lena Basin at 59°N by 15 cal ka (Mochanov 1977; Vasil'ev 2001). Their small sites and parsimonious microblade technology seems to reflect a highly mobile adaptation to scarce resources (Goebel 2002). The first permanent settlement of Beringia—including eastern Beringia—occurred within the wider context of the reoccupation of Northeast Asia during 20–15 cal ka.

The occupation of Beringia seems to have taken place at the beginning of the Lateglacial warming and transition to shrub tundra. The oldest firmly dated sites are found in the Tanana Valley of central Alaska at approximately 14.5–13.5 cal ka (Holmes 2001; C. E. Holmes, pers. comm. 2006). A site of comparable age probably is present in the lower Indigirka Basin at Berelekh (Mochanov 1977:78–79). Alternatively, previously reported dates on the lowest horizon at Ushki in central Kamchatka of ca. 17–16 cal ka (Dikov 1979) appear to be incorrect; this horizon recently has been re-dated to ca. 13 cal ka (Goebel, Waters, and Dikova 2003). And dates on large mammal bone of 18.5–15.5 cal ka probably antedate occupation of the Bluefish Caves in an upland area of the northern Yukon. The artifacts at Bluefish Caves are associated with the local shrub tundra transition (Morlan and Cinq-Mars 1982:368), which probably took place no earlier than 13 cal ka.[14]

The Lateglacial settlement of Beringia is associated with the spread of mesic shrub tundra habitat (15–13 cal ka). The base of the Birch Zone (described above) represents a pollen-stratigraphic marker for human occupation in both lowland and upland areas (Cinq-Mars 1990:21–22; Bigelow and Powers 2001). The reappearance of woody shrubs—especially dwarf birch and willow—and occasional trees that accompanied this transition may have been the critical variable in triggering human movement into Beringia at 15–14 cal ka (Guthrie 1990:276–277; Hoffecker, Powers, and Goebel 1993:48). The large quantity of burned bone among the early Beringian sites (see chapters 4–6) suggests that wood fuel often was used to ignite fresh bone (Hoffecker and Elias 2005).[15]

During the 1930s, traces of early Beringian settlement emerged in the form of two types of artifacts. In 1935, Nels Nelson reported the discovery of wedge-shaped microblade cores in central Alaska that resembled Upper Paleolithic artifacts of Northeast Asia (Nelson 1935:356). Several years later, Froelich Rainey (1939) described lanceolate projectile points—also from central Alaska—that were similar to Paleoindian points of the North American Plains. These two types of artifacts—each of which is associated with other diagnostic types—dominated discussion of Beringian archaeology in the decades that followed (figure 1.7).

More recent research provides a better understanding of how these characteristic artifacts and assemblages fit into the broader picture of Beringian settlement. The early occupation levels in the Tanana Valley reveal that the oldest firmly documented artifacts in Beringia include wedge-shaped microblade cores and microblades similar to those found in the interior of Northeast Asia during 20–15 cal ka (West 1996a;

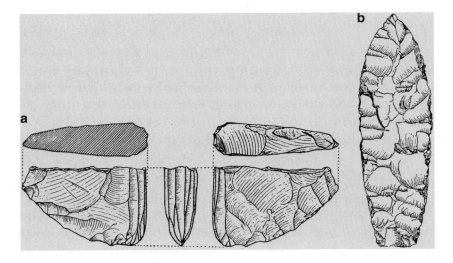

Figure 1.7 The yin and yang of Beringian archaeology: (a) microblade core similar to cores found in late Upper Paleolithic sites of Siberia (from Nelson 1937:271, fig. 16), and (b) lanceolate projectile point similar to points found in late Paleoindian sites of the Plains (from Rainey 1940:306 fig. 16); both artifacts were recovered in central Alaska during the 1930s, and each is diagnostic of the two most clearly defined archaeological complexes in Beringia (reproduced by permission of the Society for American Archaeology from *American Antiquity*, vol. 2, no. 4, 1937, and vol. 5, no. 4, 1940)

Holmes 2001). Moreover, it is clear that this technology is present in both interior and coastal areas after the Lateglacial. Despite the presence of some fluted forms in northern Alaska (Reanier 1995),[16] the lanceolate point technology appears to postdate the Lateglacial altogether (after 12.8 cal ka) and represents a long-suspected northward movement of late Paleoindian technology—apparently tied to the expansion of bison habitat during the Younger Dryas (Dumond 2001; Mann et al. 2001).

After 1960, some Beringian sites yielded new types of artifacts that lacked close resemblance to those of either interior Northeast Asia or the North American Plains. Along with microblades, small teardrop-shaped points were recovered from Healy Lake in the Upper Tanana Valley (Cook 1969), while stemmed bifacial points were found in the lowest level at Ushki on the Kamchatka Peninsula (Dikov 1977). Assemblages containing small bifacial points were excavated at sites in the Nenana Valley during the 1970s and 1980s (Powers and Hoffecker 1989), and several similar assemblages have been found in other parts of Beringia (Hamilton and Goebel 1999; Easton et al. 2004). Bifacial points also may be present in northern Alaska at this time (Rasic and Gal 2000). Many of these occupations date to the Lateglacial, and they suggest that, along with Northeast Asian microblade technology, a unique and indigenous industry emerged in Beringia before the Younger Dryas.

Whatever the character of the technology, the Beringian economy of the Lateglacial reflected a shrub tundra adaptation. Postglacial taxa such as elk and sheep dominate the large mammal remains in these sites, which also exhibit an emphasis on small mammals and birds, especially waterfowl. The pattern was noted at Healy Lake by John Cook (1969:240–243), and subsequent excavation and analysis of other sites in both western and eastern Beringia revealed a similar pattern (e.g., Mochanov 1977:79; Dikov 1996:245; Yesner 2001:321–322). Although fragments of mammoth ivory have been found at several sites, they may have been scavenged from older contexts (e.g., Yesner 2001:323). Most of the large grazers of the LGM steppe-tundra were scarce or absent by 14–13 cal ka (Guthrie 1990:285–287).

Beringia came to an end during the Younger Dryas when rising sea levels flooded the remaining portion of the land bridge between Chukotka and Alaska (12.8–11.3 cal ka). Paradoxically, inundation of the

Bering Strait took place at a time when climates had become cooler and drier (Bigelow and Edwards 2001; Elias 2001a). Both types of artifact assemblages first identified in the 1930s are present during this interval, and they are typically deposited within or below a tundra soil that formed during the subsequent warm period (see chapter 6). The lanceolate point assemblages, as already noted, apparently represent a northward intrusion of Paleoindian technology related to bison hunting and expanded steppic habitat (Dixon 1999; Dumond 2001:201–202). They seem to be confined to eastern Beringia, reflecting the absence of the land bridge.

Assemblages containing wedge-shaped microblade cores and microblades are found on both sides of the Bering Strait during the Younger Dryas. They are widely assumed to be related to the earlier Beringian industry, although they might reflect a more direct connection to the Northeast Asian sites (i.e., new movement of peoples from the Lena Basin). On the Kamchatka Peninsula, the microblade industry of the Younger Dryas is associated with two significant technological innovations: winter houses with entrance tunnels and domesticated dogs (Dikov 1979:57–60).

2

Beringian Landscapes

It is difficult to characterize the modern regions of Siberia, Alaska, and the Yukon that were formerly part of Beringia, simply because this enormous region is so environmentally diverse. Some of the southern, coastal regions experience a relatively narrow range of temperatures today, with winter temperatures that rarely dip far below freezing and summer temperatures that rarely exceed 10°C. In contrast, the interior regions of Siberia, Alaska, and the Yukon Territory experience some of the coldest winters in the world, yet summer temperatures often exceed 30°C. Following is a brief regional overview of the modern environments of the Beringian region.

A Geographic Sketch of Beringia

WESTERN BERINGIA

Northeastern Siberia contains a number of mountain ranges. From west to east, the principal ranges are the Verkhoyansk, Cherskii, and

Kolyma Mountains. The Verkhoyansk Range runs parallel and east of the Lena River, forming a great arc that begins at the Sea of Okhotsk and ends at the Laptev Sea. This great range feeds hundreds of streams that flow into the Lena as it moves northward. The Cherskii Range lies to the east of the Verkhoyansk Range and contains the region's highest peak: Pobeda (5,147 meters). Further east, the Kolyma Mountains stretch all the way to Chukotka.

One of the principal differences between western and eastern Beringia is that the former region is topographically more complex and rugged than the latter (Brigham-Grette et al. 2004). The northward-flowing river basins of the Lena, Indigirka, and Kolyma are broken by steep mountain ranges attaining heights of 2,000–3,000 meters, as well as broad uplands of 1,000–2,000 meters elevation. Broad tectonic depressions are also found in the southern parts of western Beringia, along the coast of the Sea of Okhotsk and the Anadyr River Basin.

Most of our knowledge of regional climate comes from meteorological stations in towns and villages situated along these and other rivers.[1] The winter climatic amelioration of maritime environments is clearly seen along the Pacific coast of Siberia, where mean January temperatures (TMIN) range from −4 in the south to −23°C in the north (figure 2.1). The maritime effect is much less noticeable along the coast of the Arctic Ocean (the East Siberian and Laptev Seas), because sea ice covers the coastal waters throughout the winter months, cutting off the warming effect of sea water on adjacent coastal lands. Hence TMIN along the northern coast of northeastern Siberia ranges from −27 to −37°C. The maritime influence acts to reduce summer temperatures in coastal Siberia. For instance, in Chukotka, TMAX (mean July temperature) ranges from about 4 to 8°C (figure 2.2).

Interior regions of northeastern Siberia are some of the coldest parts of the Northern Hemisphere, with TMIN values as low as −50°C. The coldest temperature ever recorded outside of Antarctica was −68°C at Verkhoyansk, in 1895. The interior regions experience the extremes of continental climate. Summer temperatures here average 15–19°C (figure 2.2), so there can be a 60–70°C temperature difference between summer and winter.

Mean annual temperatures in most of northeast Siberia are well below freezing, so the vast majority of the landscape is in the continuous permafrost zone (figure 2.3).[2] Along the southern coastal regions the

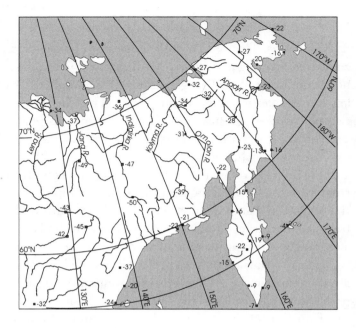

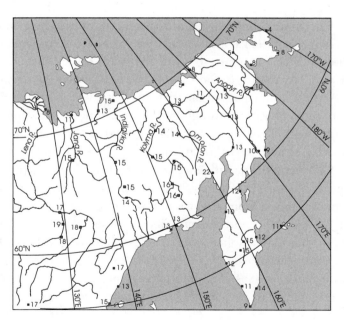

Figure 2.1 (above) The western Beringian region, showing modern mean January temperatures (°C, TMIN) (data from *Reference Book of Climate of the USSR,* 1966–1970)
Figure 2.2 (below) The western Beringian region, showing modern mean July temperatures (°C, TMAX) (data from *Reference Book of Climate of the USSR,* 1966–1970)

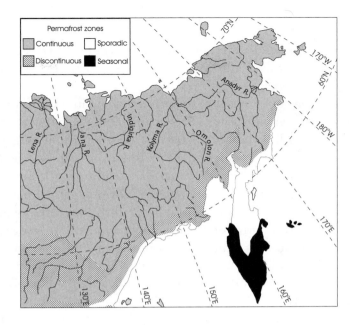

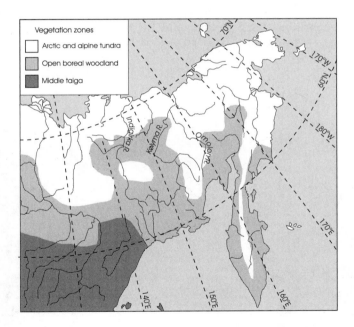

Figure 2.3 (above) The western Beringian region, showing modern permafrost zones (data from National Snow and Ice Data Center, Boulder, Colorado)
Figure 2.4 (below) The western Beringian region, showing modern vegetation zones (data from Adams and Faure 1997)

permafrost becomes discontinuous, or patchy. In northern Kamchatka the permafrost terrain is sporadic, and in the southern half, the ground only freezes seasonally.

Precipitation varies considerably in amount and in seasonal distribution across the regions that made up western Beringia, but most regions receive relatively little moisture. Some of the southernmost regions receive as much as 800 millimeters of moisture per year, but the upper reaches of the Indigirka and Kolyma Rivers receive as little as 200 millimeters of precipitation per year (Yang and Ohata 2001). As in northern Alaska, there is standing water in the form of shallow lakes, bogs, and muskeg in much of northeastern Siberia, but this is due to low evaporation rates (from the low summer temperatures) and the presence of permafrost, both of which inhibit drainage of moisture into the soil.

The modern vegetation of the regions that made up western Beringia (figure 2.4) falls within three major types: moist arctic tundra in the north, open boreal woodland in the center, and the middle taiga zone (boreal forest) in the south (Adams and Faure 1997). Within these zones there is enormous variety, as would be expected from regions that encompass millions of square kilometers of land. The arctic tundra

Figure 2.5 Pinus pumila woodland in Kamchatka (photograph courtesy of P. Krestov)

Figure 2.6 Larch woodland in the Upper Kolyma Basin (photograph courtesy of P. Krestov)

vegetation is dominated by dwarf shrubs, herbs, mosses, and lichens. However, the composition of tundra plant communities varies with local soil and climatic conditions. The tundra of Arctic coastal regions is dominated by sedges (*Carex* spp.) and by heath tundra communities (including *Eriophorum vaginatum, Vaccinium uliginosum,* and *Betula exilis*) and leafy mosses. Sheltered valleys and depressions support small stands of shrub alder, dwarf cedar, and sedge, as well as ground cover including low-bush cranberries and blueberries.

The open boreal woodland zone, as the name implies, is an open system, with trees widely spaced on the landscape. The dominant trees are conifers, such as larch and spruce. In the northern parts of this zone, botanists describe the vegetation as forest tundra, because of the mixture of trees with tundra understory vegetation. In many regions, the northern limit of trees is far from being a sharp boundary. In the eastern parts of this zone, the dominant vegetation consists of dwarf trees, such as dwarf Siberian pine (*Pinus pumila*) and alder (*Duschekia fruticosa*) (Krestov 2003). This vegetation (figure 2.5) is widely distributed in the Anadyr and Penzhina River Basins, in the Koriakskiy Mountains, and in both the northern half and the southernmost part of Kamchatka.

The middle of the Kamchatka Peninsula is clothed in forests of larch and spruce (*Picea ajanensis*), which are protected by mountain ranges from the influence of Okhotian and Pacific Ocean air masses. The dwarf-forest zone is not found in the Kolyma River Basin. Arctic tundra covers the northern parts of the Kolyma Basin, giving way to taiga forests dominated by larch, to the south (figure 2.6). B. P. Kolesnikov (1955) described a vegetation characteristic of this region, which he called "Beringian woodland." This is typified by patches of tundra vegetation, growing in disturbed sites and natural openings in the forest.[3] Larch (*Larix dahurica*) is the most stable component of the regional vegetation, and it increases in dominance toward inland regions. In river valleys and on the southern slopes, larch stands can form small forest patches (Qian et al. 2003). In the river valleys, small forest stands dominated by poplar (*Populus suaveolens*) and chosenia (*Chosenia arbutifolia*) occur in combination with wet grass meadows. Box 2.1 defines the various tundra types; figure 2.7 illustrates dwarf birch and willow.

Larch trees dominate in the north. The middle taiga region to the south also contains stands of fir (*Abies sibirica*) and pine (*Pinus pumila*). About 90 percent of the forest regions of Yakutia (the western half of western Beringia) are covered by larch.

CENTRAL BERINGIA: THE BERING LAND BRIDGE

The shallow continental shelf regions of the Bering and Chukchi Seas (figure 2.8) formed a broad connection between northeastern Siberia and western Alaska during Pleistocene glaciations. This has been determined from several lines of research. First, there is evidence from many parts of the world that sea level was lowered during Pleistocene glaciations. For instance, during the last glaciation, global sea level fell 120–130 meters below modern sea level (Bard et al. 1990). This sea-level depression was brought about because part of the water that would normally contribute to filling the world's seas was caught up in continental ice sheets and glaciers. The continental shelves between Alaska and Siberia are unusually shallow, so it follows that they were above sea level whenever sea level was substantially lower. As can be seen in figure 2.8, the entire sea shelf region between northeastern Siberia and Alaska is less than 100 meters below modern sea level. At the end of the last glaciation, rising sea level progressively flooded these shelf regions. The land bridge was flooded

Box 2.1
Classification of Tundra Types

Alpine tundra: Vegetation growing above the elevation limit of trees on mountains; in the Alaska Range, for instance, dwarf scrub communities grow on well-drained, windswept sites, and more protected slopes provide moist-to-mesic sites that support low or tall scrub communities

Arctic tundra: Vegetation growing beyond the latitudinal limit of trees (i.e., growing north of the treeline in the Arctic)

Dry herbaceous tundra: Vegetation dominated by herbs (grasses, sedges, and other nonwoody tundra plants) growing in relatively dry soils, such as uplands with sandy substrates

Forest tundra: Mixture of often widely spaced trees with an understory of tundra vegetation; this type of vegetation is more common in western Beringian regions than in eastern Beringia

Heath tundra: Vegetation dominated by heather (*Cassiope*) and including such species as cotton grass (*Eriophorum vaginatum*), bog blueberry (*Vaccinium uliginosum*), mountain-cranberry (*V. vitis-idaea*), dwarf birch (*Betula exilis*), alpine bearberry (*Arctostaphylos alpina*), crowberry (*Empetrum nigrum*), and narrow-leaf Labrador-tea (*Ledum decumbens*) (Gallant et al. 1995)

Herbaceous (or graminoid) tundra: Vegetation dominated by grasses, sedges, and other herbs (nonwoody plants)

Mesic tundra: Vegetation growing in mesic (medium-moisture) conditions

Mesic herbaceous tundra: Herbaceous vegetation growing in mesic (medium-moisture) conditions; in northern Alaska, this vegetation is dominated by tussock-forming sedges such as cotton grass and Bigelow's sedge (*Carex bigelowii*). Low shrubs such as dwarf arctic birch (*Betula nana*), crowberry, narrow-leaf Labrador-tea, and mountain-cranberry often occur and may co-dominate with sedges. Mosses (e.g., *Hylocomium splendens* and *Sphagnum* spp.) and lichens (e.g., *Cetraria cucullata*, *Cladonia* spp., and *Cladina rangiferina*) are common between tussocks (Gallant et al. 1995).

Moist nonacidic tundra: Growing on calcium-rich, fine-grained soils with relatively high pH. The dominant plants are sedges (e.g., *Carex bigelowii*, *C. membranacea*, and *Eriophorum triste*), prostrate and dwarf shrubs (e.g., *Dryas integrifolia*, *Salix reticulata*, *S. arctica*, *S. lanata*, and *S. glauca*), and numerous minerotrophic mosses (e.g., *Tomentypnum nitens*, *Distichium capillaceum*, and *Ditrichum flexicaule*). It also has many forb species (Walker et al. 2001).

(Box 2.1 continued)

Shrub tundra: Dominated by dwarf shrubs, such as dwarf birch, dwarf willow, narrow-leaf Labrador-tea, mountain-cranberry, and crowberry; this type of vegetation requires mesic (medium-moisture) to moist conditions

Steppe-tundra: Mixture of plant species found today in steppe regions (mid-latitude to high-latitude grasslands) and arctic tundra regions. This once-widespread vegetation type is now restricted to relict patches on south-facing slopes of mountains in both eastern and western Beringia. The vegetation included a wide variety of herbs and dwarf shrubs. The herbs included tufted grasses, sedges, and tufted sedges. The tufted grasses included several species of fescue (*Festuca*); grasses, including *Poa botryoides, P. stepposa, P. arctostepporum, P. glauca, Calamagrostis purpurascens,* and *Helictotrichon krylovii; Koeleria cristata* and *K. asiatica*; and wheatgrasses of the genus *Elytrigia*. The sedges included true steppe species such as *Carex duriuscula*; meadow-steppe species such as *C. obtusata* and *C. rupestris*; and tufted sedges such as *Carex pediformis, C. filifolia, C. rossii, C. aenea, C. petasata*, and others. Dry-adapted and cold-dry-adapted dicot plants included several species of sage (*Artemisia*), cinquefoil (*Potentilla*), milk-vetch (*Astragalus*), and locoweed (*Oxytropis*) (Yurtsev 2001).

Subarctic tundra: Subarctic regions where climatic factors (cold summers, high winds, etc.) prevent the establishment and growth of trees: mesic graminoid herbaceous communities and low scrub communities occupy extensive areas on hills and lower mountain slopes; wet meadows and bogs are dominant on saturated soils

Tussock tundra: Vegetation growing in the form of tussocks, compact mounds of grasses or sedges, held together by root masses, growing in mesic-to-moist environments (see description of mesic herbaceous tundra for a list of typical tussock plant species)

Wet herbaceous tundra: Vegetation dominated by moisture-tolerant sedges and grasses, often with standing pools of shallow water. On the North Slope of Alaska, wet herbaceous tundra is dominated by sedge communities with *Carex aquatilis* and *Eriophorum angustifolium*. Grass communities are generally dominated by *Dupontia fischeri* and *Alopecurus alpinus*, but *Arctophila fulva* dominates where surface water is 15 to 200 cm deep (Gallant et al. 1995).

Figure 2.7 Dwarf willow (left) and birch (right) (from E. Hultén, *Flora of Alaska and Neighboring Territories,* copyright © 1968 by the Board of Trustees of the Leland Stanford Jr. University, renewed 1995)

sometime shortly after 12.9 cal ka (12,900 calibrated years) (Elias et al. 1996), based on the youngest radiocarbon ages from terrestrial plant remains, taken from Chukchi Sea sediment cores. These sediments were buried by marine sands that swept over the land bridge when sea level rose and saltwater flooded the continental shelves.

EASTERN BERINGIA: ALASKA AND THE YUKON TERRITORY

The Arctic coastal plain of northern Alaska slopes gently northward toward the Arctic Ocean. The permafrost layer extends down more than 300 meters in many places in this region. Although the coastal

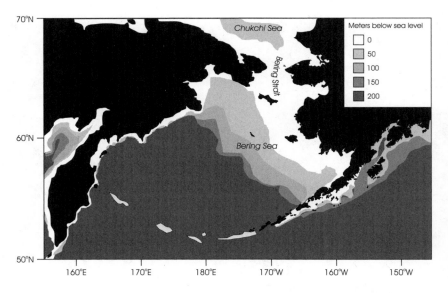

Figure 2.8 The Bering and Chukchi sea shelf regions between Siberia and Alaska, showing the bathymetry

plain receives little precipitation, the landscape is dotted with an abundance of shallow lakes, because soil drainage is inhibited by permafrost and little water evaporates in the cold arctic climate.

South of the coastal plain, the Arctic foothills zone consists of northward-sloping foothills, extending from the north slope of the Brooks Range to the coastal plain. Elevations in the foothills zone range from 180 meters in the north to 1,100 meters in the south. Most of the rivers and streams that course through the foothills down to the coastal plain flow consistently northward, except for the east-flowing upper portion of the Colville River.

The Brooks Range, the Arctic foothills region, and the De Long and Baird Mountains form an east-west trending mountain barrier that separates the coastal plain in the north from the interior regions to the south.[4] The Brooks Range contains the highest mountains north of the Arctic Circle. This range forms the divide between the northward-flowing waters heading to the Arctic Ocean, the westward-flowing waters heading to Kotzebue Sound of the Chukchi Sea, and the southward-flowing waters of the Yukon River drainage that empty into the Bering Sea.

South of the Brooks Range is the vast Alaskan interior, bounded on the south by the Alaska Range. The Yukon and Kuskokwim Rivers form the principal drainages for this large region. Both rivers empty into the Bering Sea. Most of this region is made up of hills, plateaus, and rolling uplands, with elevations ranging from 600 to 1,500 meters. The Yukon Flats region is an exception to this varied topography. Within the interior region are several highly dissected upland regions. The mountains range in elevation from 1,220 to 1,830 meters. There are no active glaciers in this region. It is underlain by "continuous" permafrost only at high elevations (figure 2.9).

The westernmost extent of the Alaskan interior region is the Seward Peninsula. This peninsula consists of broad hills and ridges averaging about 600 meters, but in the more rugged mountains a few peaks rise above 900 meters. The highest point on the Seward Peninsula is Mount Osborn, at 1,440 meters. The entire peninsula is underlain by permafrost.

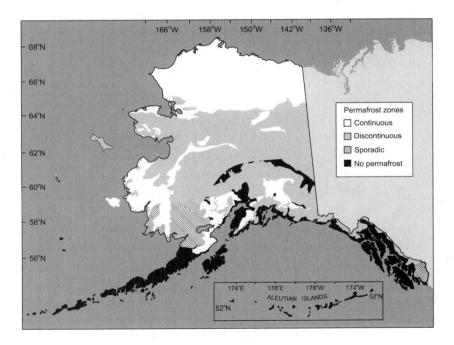

Figure 2.9 Alaska, showing modern permafrost zones (data from Brown et al. 1997)

The mountains in the southern half of Alaska are the most massive and spectacular mountain system in North America. They form the northwestern sector of the Pacific Mountain system that follows the west coast of the Americas from Alaska to Central America. As in other parts of this Pacific Mountain system, the southern Cordillera of Alaska is split into two belts of mountains, separated by intervening lowlands. In Alaska, the northern mountain belt is the Alaska Range. The Alaska Range forms a 1,000-kilometer arc that roughly parallels the Pacific coast of Alaska. The range is extremely rugged and heavily glaciated.[5] Four great mountain masses dominate the Alaska Range, divided by a number of low passes and river valleys. Great valley glaciers radiate from these mountain masses, except in the relatively low eastern sectors, where the few glaciers are small. The glaciers on the south side of the range contain substantially more ice than those on the north, because the southern glaciers receive more Pacific moisture than the northern glaciers, due to the rain shadow effect of the tall mountains. The Alaska Range is the drainage divide for rivers flowing north into the Yukon River system and those flowing south into the Gulf of Alaska.

The western edge of the Alaska Range converges with the Aleutian Range and the Aleutian Islands. These islands are a chain of volcanic peaks that lie on the crest of a huge submarine ridge rising 3,350 meters above the sea floor. The Aleutian ridge is 2,400 kilometers long and 30–100 kilometers wide; it separates the Bering Sea from the Pacific Ocean. The archipelago stretches in a long arc from the Alaska Peninsula in the northeast to Attu Island in the southwest. The eastern margin of the Alaska Range converges with the Boundary Ranges and Coastal Foothills of the Alaskan Panhandle.

The southern mountain belt of the southern Cordillera includes the Kodiak, Kenai, and Chugach Mountains in the west, and the Wrangell, St. Elias, Chilkat-Baranof, and Prince of Wales Mountains in the east. The mountains of Kodiak Island, the Kenai Peninsula, and the Chugach Mountains combine to form a rugged barrier along the northeast margin of the Pacific Ocean. The average elevation of the Kenai Mountains is 900–1,500 meters. The Chugach Mountains are somewhat higher, averaging 2,100–2,440 meters. These mountain regions are extremely rugged and heavily glaciated. The Kenai Mountains contain the Sargent and Harding Ice Fields, and the eastern Chugach Mountains contain the Bagley Ice Field.

The St. Elias Mountains, one of the highest coastal mountain groups in the world, straddle the border between Alaska and Canada.[6] The chain is about 320 kilometers long and has a maximum width of about 160 kilometers. These massive, block-like mountains form an array of narrow ridges and spectacular peaks averaging 2,400–3,000 meters. There are also many ice-covered peaks higher than 4,200 meters in the range. Mount Logan is the highest peak in the range and is Canada's highest mountain, at 5,951 meters. Mount St. Elias at 5,488 meters straddles the Yukon-Alaska border.

The Wrangell Mountains, covering an area of roughly 18,000 square kilometers, have high, glaciated massifs that rise more than 3,000 meters above the Copper River Basin. The Copper River is the only large river draining south-central Alaska. Ice caps cover most of the highest mountains and feed large valley glaciers.

The lowlands that separate the southern belt from the northern belt include, from west to east, the Cook Inlet-Susitna and Copper River Lowlands, and the Kupreanof Lowlands of southeastern Alaska. These lowland regions are connected by narrow, broken valleys along the southern margin of the Talkeetna Mountains and along the north side of the Wrangell and St. Elias Mountains.

Alaskan climate is quite variable, given its high-latitude position.[7] Temperatures generally decline northward in Alaska, and they decrease with increasing elevation. The maritime climate of southern Alaska buffers the extremes of seasonal temperatures seen elsewhere. For instance, the difference between mean July and mean January temperatures at Sitka is only 12°C. The difference between mean January and mean July temperatures at Fairbanks is 41°C. East-central Alaska experiences the warmest summers and some of the coldest winters of the state, so the differences between seasons are rather extreme.

Winters in the interior and northern regions of Alaska are certainly cold, but not as cold as those of the same latitudes in Siberia. This is because Alaska benefits from southern winds that circulate around a low-pressure system in the North Pacific, called the Aleutian low. These winds bring relatively warm North Pacific air masses into Alaska. During July, atmospheric circulation patterns generally shift northward as more radiation and warm high-pressure systems become more dominant. The jetstream is most prominent over Alaska during summer, bringing summer rains at most locations as storms travel from west to east. July

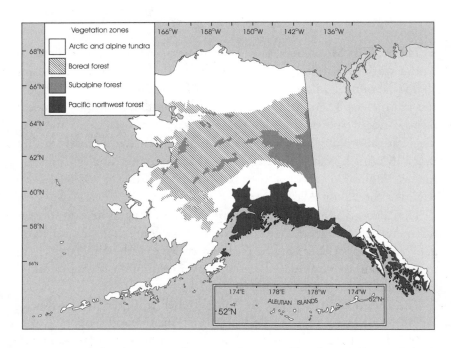

Figure 2.10 Alaska, showing modern vegetation zones (data from Gallant et al. 1995)

precipitation varies from 50 to 100 millimeters, with higher values generally southward, but some values in high mountain ranges exceed 100 millimeters. Winter precipitation is the lowest of the annual cycle throughout most of Alaska. For instance, many locations average less than 25 millimeters in January, except for south coastal Alaska. Mean annual temperatures range from a low of –13°C at Barrow to 7°C at Sitka. Likewise, mean annual precipitation is highly variable. It ranges from 5,800 millimeters at Little Port Walter, south of Sitka on Baranov Island, to only 114 millimeters at Barrow. Alaska has almost as much variation in annual precipitation as the rest of the United States combined. In between the two extremes, most Alaskan regions receive 250–400 millimeters of precipitation per year. In the interior of Alaska, precipitation mostly falls as snow from September to May, although spring rains are not uncommon.

Alaska is home to more than 1,550 species of plants, according to Eric Hultén (1968). Alaska enjoys a great floral diversity because its wide range of environments provides a wide variety of habitats for plants. The major plant communities of Alaska fall into about twenty major types. These communities, along with soil characteristics, climate, and topographic features, have been described as ecoregions (figure 2.10) (Gallant et al. 1995).

NORTHERN ALASKA

The three ecoregions in northern Alaska represent arctic ecosystems. The northern limit of trees is found on the southern flanks of the Brooks Range. At higher elevations and to the north, alpine tundra grades into arctic tundra. There are patches of alpine tundra in mountain ranges to the south, and the subarctic tundra of the Seward Peninsula and southwestern Alaska has many species in common with the arctic tundra, so only a few plant species are unique to the arctic regions of Alaska. However, the arctic plant communities are unique, because they grow in environments unlike those found anywhere else in Alaska.

The Arctic coastal plain is a relatively flat, poorly drained plain that rises very gradually from sea level to the Arctic foothills to the south. The

Figure 2.11 Wet herbaceous tundra vegetation near Barrow, Alaska (photograph by SAE)

coastal plain has very low temperatures and very low annual precipitation, and the entire area is underlain by thick permafrost.[8] Although July and August are generally frost-free, freezing temperatures can occur in any month of the year. Although the region receives an average of only about 140 millimeters of precipitation annually, wet soil conditions are most typical; these soils support wet herbaceous tundra dominated by sedges or grasses. Dryer soil conditions occur on slight rises, and dwarf scrub communities grow in these better-drained areas (figure 2.11).

The Arctic foothills region consists of rolling hills and plateaus that stretch from the coastal plain in the north to the Brooks Range in the south. The hills and valleys have better drainage than the coastal plain, and there are fewer lakes. The region is underlain by thick permafrost, and many ice-related surface features are present. In winter, the foothills are somewhat warmer than the adjacent regions to the north and south, and this region is warmed somewhat by the Chukchi Sea to the west. Vegetation over most of the region consists of mesic herbaceous tundra and dwarf scrub tundra.[9]

Figure 2.12 Sparse vegetation cove in the De Long Mountains (photograph by SAE)

The Brooks Range ecoregion forms an arc across northern Alaska, from the Canadian border to within 100 kilometers of the Chukchi Sea. The climate of the Brooks Range is extremely cold. Winter temperatures drop to –50°C, and July temperatures average less than 10°C, although freezing occurs during all the summer months. Continuous, thick permafrost underlies the region. Rubble and exposed bedrock cover the mountain slopes. Vegetation cover is sparse (figure 2.12), because of continual erosion of mountain slopes, shallow soils, high winds, and harsh climate. Most high-elevation regions are essentially devoid of vegetation. Dwarf scrub and mesic herbaceous tundra communities are generally limited to valleys and lower hill slopes.

INTERIOR ALASKA

Between the Alaska Range to the south and the Brooks Range to the north lies the vast interior region of Alaska, mostly covered by boreal forests dominated by white and black spruce (*Picea mariana*), with stands of balsam poplar (*Populus balsamifera*), paper birch (*Betula*

Figure 2.13 Coniferous forest, woodland, and alpine tundra on the northern slope of the Alaska Range (photograph by SAE)

papyrifera), and quaking aspen (*Populus tremuloides*). General patterns of plant communities are largely controlled by elevation, drainage patterns, and permafrost conditions.[10]

The Interior Highlands region is a mixture of rounded, low mountains and rugged peaks. The highlands primarily sustain dwarf scrub vegetation and open spruce stands, though herbaceous communities occur in poorly drained areas (figure 2.13). The region has a continental climate. The northern part of this region is underlain by continuous permafrost, and the central and southern parts are underlain by discontinuous permafrost. The highest mountain slopes are devoid of vegetation. Dwarf scrub communities are common in windswept sites. Lower elevations are generally more protected from wind and have a denser vegetation cover that can include open coniferous forests and woodlands. Open coniferous forests and woodlands are often dominated by white spruce or a mixture of white and black spruce.

Much of the high country of the Alaska Range is covered by ice fields and glaciers. Even the ice-free alpine regions are mostly devoid of vegetation. The climate is extremely cold at high elevations, and the soils are generally shallow over bedrock. The mountain ridges of the Alaska Range are separated by broad, glacially carved valleys.

Dwarf scrub communities are most common where vegetation does occur, growing on well-drained, windswept sites. Below the high mountain summits, slopes more protected from wind provide moist to mesic habitats that support low or tall scrub communities. Open coniferous forests and woodlands occur on well-drained sites in some valleys and on lower hill slopes. These are dominated by white spruce or a mixture of white and black spruce.

SEWARD PENINSULA

The predominantly treeless region of the Seward Peninsula is surrounded by water on three sides, but the winters are nonetheless long and harsh and the summers are short and cool. The peninsula has narrow strips of coastal lowlands that grade into extensive uplands of broad hills and flat valleys. Small, isolated groups of rugged mountains up to 1,400 meters elevation occur in a few locations. Permafrost is continuous throughout the region, and periglacial features are common. The lowlands are dotted with thaw lakes. Mesic herbaceous communities and low scrub

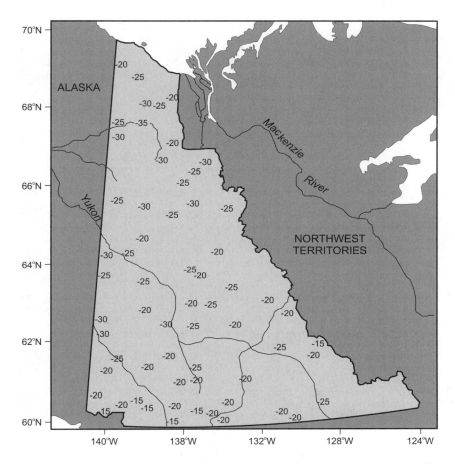

Figure 2.14 The Yukon Territory, showing mean January temperatures (°C, TMIN) (data from Wahl et al. 1987)

communities cover extensive areas on hills and lower mountain slopes. Wet herbaceous communities grow on poorly drained lowlands. Tall scrub vegetation occurs along streams and flood plains. The highlands are either barren of vegetation or support dwarf scrub communities. Wet herbaceous communities consist of sedges and grasses.

The lowlands of southwestern Alaska include the coastal plains of the Kotzebue Sound area and the Yukon and Kuskokwim River delta area. This flat, poorly drained region has many lakes, and the deltas of the huge Kuskokwim and Yukon Rivers are important features on the landscape. The Pribilof Islands, Nunivak Island, and part of St. Lawrence Island are also in this ecoregion. Not surprisingly, the

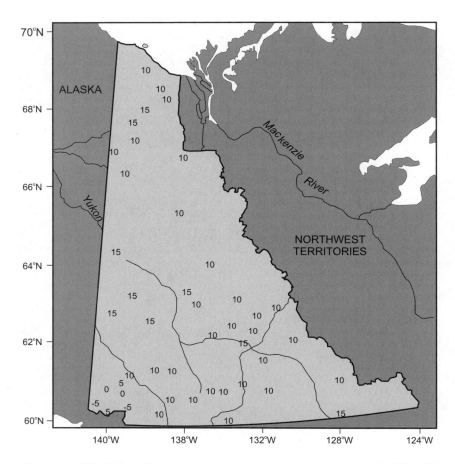

Figure 2.15 The Yukon Territory, showing mean July temperatures (°C, TMAX) (data from Wahl et al. 1987)

regional soils are wet and the permafrost table is shallow. The region is affected by both maritime and continental climatic influences. Wet herbaceous communities dominate the region.[11]

YUKON TERRITORY

Unlike Alaska, which has a large interior lowland region, the Yukon Territory is more rugged, with mountains, rough plateaus, and valleys. The southern two-thirds of the territory is essentially a highland region, with most terrains above 900 meters in elevation (Scudder 1997). These highlands are part of the Cordilleran mountain system that extends southward through British Columbia and Alberta and further south in the Rocky Mountains to New Mexico. The northeastern part of the Yukon is part of the Interior Plains region of Canada. The northernmost part of the Yukon falls within the Arctic coastal plain region described above for northern Alaska. The British Mountains form an eastern extension of the Alaskan Brooks Range. Steep-sided, rugged peaks in these mountains reach elevations up to 1,680 meters. The Richardson Mountains that lie along the northeastern edge of the territory are more highly eroded and form more gentle slopes. The central region of the territory,

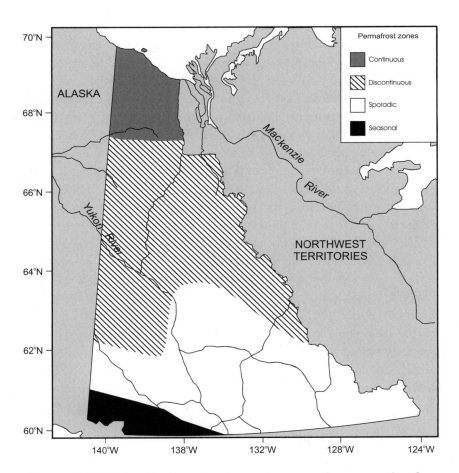

Figure 2.16 The Yukon Territory, showing modern permafrost zones (data from Wahl et al. 1987)

the Yukon Plateau, forms the headwaters of the Yukon River drainage. This region is a topographically complex highland, including five mountain ranges. The southwestern Yukon contains the Icefield and Kluane Ranges and the St. Elias Mountains. These rugged highlands have many peaks over 2,000 meters, including Mount Logan. Permanent snow and ice cover all but the highest peaks in parts of these mountains.

Yukon climate is highly variable but typified by long, cold winters and short, warm summers (figures 2.14, 2.15).[12] Permanently frozen ground occurs throughout most of the Yukon. Continuous permafrost occurs throughout the northern part of the territory. Most of the central region is in the zone of discontinuous permafrost, and the southern third of the territory is in the zone of sporadic permafrost (Wahl et al. 1987). Only the lower elevations of the southwest corner of the Yukon lie outside the permafrost zone and experience seasonally frozen ground (figure 2.16).

The plant communities of the Yukon fall into four broad categories: arctic and alpine tundra, subalpine forest, and boreal forest (figure 2.17). Arctic tundra dominates the northern coastal plain region,[13] and alpine tundra regions are extensive in the Yukon.[14] The subalpine forests of the Yukon are described as taiga by some authors (Scudder 1997) in keeping with the Siberian designation of taiga as open boreal woodland. These woodlands occur on the lower slopes of the various mountain ranges that dominate the central regions of the Yukon, and they are scattered in the series of mountain groups in the southern half of the territory.[15]

The boreal forest region that dominates the southeastern Yukon is composed of stands of black and white spruces. Patches of boreal forest extend up to elevations of 800 meters along river valleys in the central Yukon (Kojima and Brooke 1986). White spruce is the dominant upland tree species. Disturbed sites are invaded by black spruce, quaking aspen, balsam poplar, paper birch, and larch. Two other tree species that are more common in the southern parts of the Cordillera can also be found here: lodgepole pine (*Pinus contorta*) and subalpine fir (*Abies lasciocarpa*). Lodgepole pine and quaking aspen come into recently burned sites; fire is an important element in the ecology of boreal forest stands.

THE BERING LAND BRIDGE

During the Pleistocene, the alternating cycles of inundation and exposure of the land bridge have largely been dictated by changes in

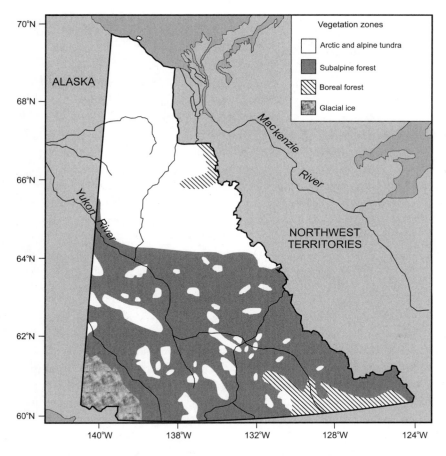

Figure 2.17 The Yukon Territory, showing modern vegetation zones (data from Wahl et al. 1987)

global sea level. When global sea levels have fallen because of the onset of glaciation, the land bridge has been exposed. The land bridge emerged during major glaciations when sea levels fell by 100–135 meters (Hopkins 1973; Clark and Mix 2002) and has been a major migration route for plants and animals—the only connection between continents in the northern high latitudes. When sea levels have risen because of deglaciation, the land bridge has been submerged. Because interglacial periods comprise only about 10 percent of the last 2.7 million years, the land connection between the two continents was only occasionally broken.

Box 2.2
The Radiocarbon Timescale

In this book, we present radiocarbon ages in the form of calibrated years before present (BP). This requires some explanation. During the past 20 years, scientists have been able to check the accuracy of radiocarbon ages through the dating of tree rings and annually layered lake and marine sediments. By dating sequences such as these that can be counted back through time, it has been possible to determine the accuracy of radiocarbon ages. As illustrated in figure 2.17, the radiocarbon time scale, when matched against calendar years (as counted by tree rings, for instance), has been shown to be nonlinear. In other words, radiocarbon years are not calendar years. Because of variations in the creation of ^{14}C atoms in the atmosphere (due to changes in solar activity), the trapping and then sudden release of ancient organic-rich sediments beneath glacial ice sheets, and other variations in Earth's carbon cycle, the radiocarbon time scale drifts above and below the actual time scale—especially during certain intervals in the past 15,000 years.

Figure 2.18 shows two different radiocarbon ages and how these ages vary considerably in their precision. In figure 2.18a, a point in radiocarbon time matches with only a narrow band of calendar time. Thus, when this ^{14}C age is calibrated against calendar years, it yields a very narrow band of time. In figure 2.18b, however, because of the drifting timescale of radiocarbon years, a radiocarbon age yields a calibrated age that spans a great deal of time. In the latter case, the time interval in question is considered a radiocarbon *plateau* time, because a radiocarbon age in this zone corresponds with three different calendar age groups. Unless one is dating tree rings or annually countable sediment layers, it is impossible to determine which of the three calendar age ranges correspond to the radiocarbon age. Therefore the only unbiased way to treat such a radiocarbon age is to say that the events associated with that age happened sometime within the broad time zone indicated by the calibration curve.

Unfortunately for Beringian archaeology, one of the largest of these radiocarbon plateaus falls just at 11,000 ^{14}C years BP (Younger Dryas Radiocarbon Plateau). Calibrated calendar years for this radiocarbon age fall between 12.8 and 12.9 cal ka (12,830–12,931 calibrated years ago). So, a single radiocarbon age encompasses more than 100 calendar years, just at one of the most critical points in the prehistory of Beringia and the peopling of the New World.

For the purposes of this book, the most important inundation of the Bering Land Bridge was the one that took place at the end of the last glaciation. The timing of this transgression has been problematic (box 2.2, figures 2.18 and 2.19).[16] As noted in chapter 1, Elias et al. (1996) had the opportunity to extract and identify terrestrial plant macrofossils from sediment cores taken from the Chukchi shelf region. Macrofossils from the youngest terrestrial sediments of the land bridge yielded ages that demonstrated that much of the land bridge remained intact until after 12.9 cal ka. The revised age for inundation of the land bridge confirmed that it was available for human movement into the New World until the end of the Pleistocene. Paleotemperature reconstructions for this interval from Alaska (Elias 2000) suggest that summer temperatures were at least as warm as they are today. However, this does not, of itself, indicate that eastern Beringian environments were altogether benign and hospitable to nomadic human hunters and their prey. As

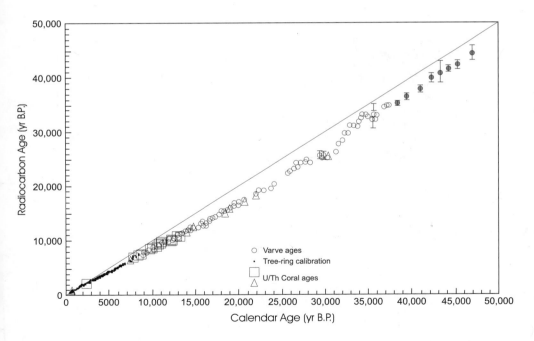

Figure 2.18 Radiocarbon time scale versus calendar years (after Kitagawa and van der Plicht 1998)

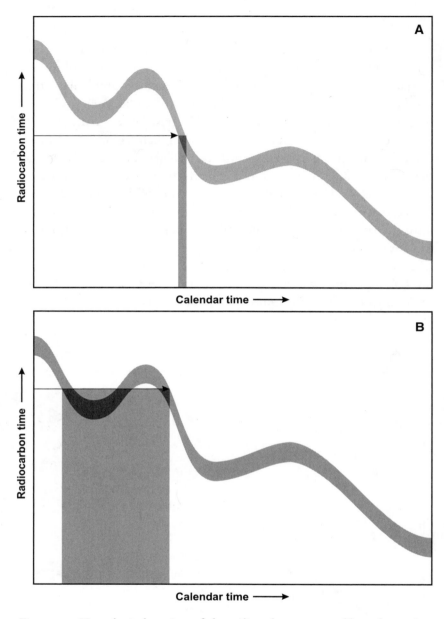

Figure 2.19 Hypothetical section of the radiocarbon versus calibrated age time scales, showing (A) a point in radiocarbon time that corresponds to a narrow segment of calibrated time and (B) a point in radiocarbon time that corresponds to a very broad segment of calibrated time because it intersects more than one point in the calibrated time scale

discussed here later, the transition from the LGM to the Holocene was a time of ecological upheaval.

The land bridge served as an important conduit for plant and animal migration between Asia and the Americas.[17] Many species of the Beringian mammal fauna ranged across both sides of the land bridge during the late Pleistocene, including several species of horse, musk oxen, and woolly mammoth. A band of mesic tundra may have existed on the land bridge (Elias et al. 1996), and as noted in chapter 1, Guthrie (2001) suggested that this mesic tundra environment may have blocked eastward migration of some dry steppe-adapted species, such as woolly rhinoceros from western Beringia. Other steppe-adapted mammals, such as saiga, entered eastern Beringia but subsequently died out in the New World (saiga is now confined to the steppes of central Asia).

What were the environments of the land bridge itself? We are still coming to terms with this question, based on a few isolated fossil records taken from sediment cores in the Bering and Chukchi Seas and from islands that once were highlands on the land bridge. The evidence gathered so far indicates the following. At the northern edge of the Bering Land Bridge, the continental shelf of the Arctic Ocean was also above sea level. The Arctic Ocean was cut off from warm Pacific water when the land bridge came into being. The Arctic Ocean froze, then remained frozen at the surface until the land bridge was inundated at the end of the Pleistocene. The northern coast of the land bridge was probably extremely cold and offered only marginal habitats for terrestrial life. The vegetation in this region was likely similar to that found in modern polar deserts of the high arctic, and—because the northern coast was icebound—probably supported few marine mammals for human hunters.

The cutting-off of Pacific waters and moist air masses by vast expanses of continental shelf at the southern end of the land bridge produced the cold and arid climate that fostered the steppe-tundra biome across Beringia. We do not have a clear picture of what the vegetation was like on much of the land bridge itself, however, because little work has been done on sampling and studying fossil pollen and macrofossils from land bridge sediments.[18] Although Elias et al. (1996), and Elias, Short, and Birks (1997) described pollen, plant macrofossil, and insect fossil assemblages from sediment cores taken from the Bering and Chukchi shelves, only one of these samples dated from the LGM. Nevertheless, all of the samples (ranging between 23 and 12.86 cal ka) yielded fossils indicative of mesic tundra conditions, including shrub tundra vegetation (dwarf birch and willow) and mesic tundra beetles. In contrast, Ager (2003) studied pollen from lake sediments on St. Michael Island in the Bering Sea and found evidence of steppe-tundra environments at that locality during 26–16 cal ka. He also cited evidence from other published pollen studies of late Pleistocene lake sediments, including a site on the Pribilof Islands (Colinvaux 1981), that suggest the presence of steppe-tundra or herb-dominated tundra environments during 25–18.2 cal ka.[19]

Reconstructing Beringian Environments

From a glaciological standpoint, Beringia was an anomaly. Nearly all other high-latitude regions of the Northern Hemisphere were covered by ice sheets during glacial periods, but the lowlands of Beringia remained ice-free, thus providing a refuge for high-latitude flora and fauna. The buildup of ice sheets at high latitudes requires two principal conditions: low temperatures and abundant moisture. Beringia, along with the rest of the arctic and subarctic regions, was certainly sufficiently cold during glacial episodes to foster the growth of glacial ice, but it was too dry. The Bering Land Bridge effectively blocked moist air masses from the North Pacific Ocean from reaching the interior regions of both eastern and western Beringia (see figure 1.1). The only substantial ice cover in Beringia during Pleistocene glaciations was in the form of mountain glaciers. The largest of these was the westernmost part of the Cordilleran Ice Sheet of western North America. This ice sheet covered the Alaska Range and extended south and west to bury the Alaska Peninsula and the continental shelf regions of Prince William Sound and the Gulf of Alaska (Mann and Hamilton 1995). Unlike the North Atlantic, much of the North Pacific Ocean probably remained free of sea ice during the LGM (Jaccard et al. 2005).

The interpretation of Beringian pollen assemblages is difficult and given to widely varying interpretations. For instance, Anderson and Lozhkin (2001) noted that the Middle Wisconsin vegetation from the Epiguruk site in northwestern Alaska has been interpreted as both *Salix*-graminoid tundra (Hamilton et al. 1993) and *Betula* shrub tundra with numerous floodplain microhabitats (Schweger 1982). Hamilton et al. (1993) concluded that the Epiguruk vegetation was sufficiently productive to support populations of Pleistocene megafauna.

There was apparently a significant difference between the climate and vegetation of western and eastern Beringia during the MIS 3 interstadial (ca. 48–28 cal ka). In western Beringia, temperatures apparently rose to the level that allowed larch forests to expand northward to virtually their modern limits (Anderson and Lozhkin 2001). As can be seen in figure 2.20, this phenomenon was time-transgressive across the various regions of western Beringia. In contrast to this, however, there is very little evidence for the expansion of coniferous forest in eastern Beringia during MIS 3. Based on fossil beetle evidence, there appears to have been different degrees of warming in the northern and southern sectors of eastern Beringia during this interstadial (Elias 2000). The arctic regions experienced mean summer temperatures as much as 1.5°C warmer than modern levels at about 35 cal ka, whereas the synchronous warming in subarctic regions remained about 2°C cooler than modern levels.

The key issue for the health of megafaunal mammal populations is not temperature but primary productivity (box 2.3): the rate at which regional vegetation converts sunlight, water, carbon dioxide, and soil minerals and nutrients into organic matter or plant biomass. The productivity of Beringian landscapes is a key element in unraveling the mystery of human settlement there—or the lack of it. In the scenarios presented by Guthrie (2001) and Zimov et al. (1995), a highly productive grass-dominated steppe-tundra supported large numbers of megafaunal grazers and their predators. The palynologists studying Pleistocene environments of Beringia generally take a more cautious approach. For instance, Anderson and Brubaker (1994) described the LGM vegetation in eastern Beringia as follows:

> We believe that the presence of minor pollen taxa with strong arctic affinities (e.g., *Oxyria, Saxifraga, Dryas, Saussurea*), the minor presence of shrub taxa (Yurtsev 1981), and the statistical similarity of Duvanny Yar spectra to modern arctic pollen assemblages (Anderson et al., 1989) argue for the presence of herb-dominated tundra rather than grassland or steppe (Anderson and Brubaker 1994:84)

Are these differences of interpretation simply a matter of semantics, or would the two types of reconstructed landscapes have been significantly different? Modern studies of plant productivity would argue for

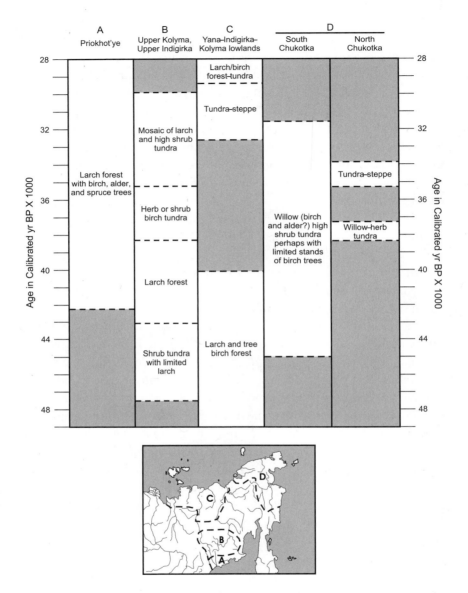

Figure 2.20 Marine Isotope Stage 3 (MIS 3) interstadial vegetation summary for western Beringia, based on pollen analysis (after Brigham-Grette et al. 2004)

Box 2.3
The Productivity Paradox Problem

In the 1980s, some palynologists studying Late Pleistocene pollen assemblages from eastern Beringia were convinced that the landscapes of the Last Glacial Maximum (LGM) were essentially polar deserts with only sparse, herbaceous ground cover. In their study of pollen from sites in the Yukon, Ritchie and L. C. Cwynar (1982:113) concluded that "the 'arctic steppe biome' never existed in eastern Beringia during the late Quaternary." This conclusion was based mainly on the low pollen productivity and species diversity seen in the Yukon pollen records of this time interval. Influx rates of herbs and sagebrush (the two key elements of the proposed steppe-tundra biome) were much lower than they are in modern pollen profiles from the Canadian prairies. Accordingly, these authors maintained that the large and diverse ungulate populations found in eastern Beringian fossil assemblages must have lived during interstadial intervals greater in age than 30 cal ka, for which no pollen records were then available.

In sharp contrast to this view, Guthrie (1982) argued that the large, diverse megafaunal mammal populations of eastern Beringia could only have survived in a highly productive steppe-tundra environment that he termed "mammoth steppe." Herein lies the "productivity paradox." One set of paleobotanists argued for a low-productivity polar desert, whereas the vertebrate paleontologists argued for a high-productivity steppe. With the hindsight provided by more than 20 additional years of fossil data collection, it seems safe to conclude that both sides of the argument were wrong, in one important way. Each side of the argument was essentially envisioning Beringia as a monolithic biome, dominated by a single type of vegetation. From an ecological point of view, this makes no sense. No region as large as eastern Beringia could ever be ecologically uniform. The topographic, edaphic, and climatic diversities of such an immense region are enormous. It now seems clear, as was argued by Bliss and Richards (1982) and Schweger (1982), that eastern Beringia (and by logical extension, western Beringia) was clothed in a mosaic of many different plant communities.

the latter case. These studies show marked productivity differences between arctic tundra and steppe vegetation.[20] Based on some models, such as Guthrie's (2001) reconstruction of mammoth steppe, it appears that a cold steppe landscape would have been able to support large numbers of grazing mammals—potential prey for human hunters.

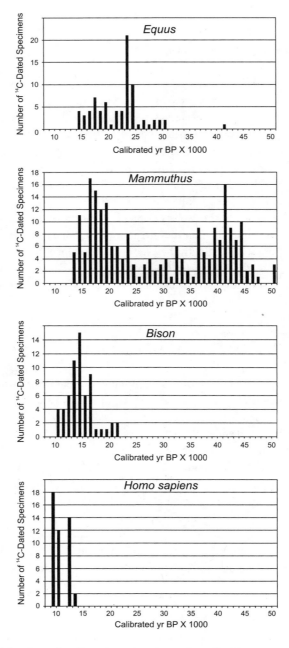

Figure 2.21 Number of radiocarbon-dated remains of various megafaunal mammals found in eastern Beringia (data from Guthrie 2006)

Herb-dominated tundra, while not exactly like modern arctic tundra, would probably have been far less productive, supporting fewer large mammals.

Anderson and Lozhkin (2001) describe steppe-tundra vegetation as dominant in both the cool phase of the Karginskii (MIS 3) interstadial and the LGM of western Beringia, so there seems to be less controversy about steppe-tundra in western Beringia. But what about eastern Beringia? If the palynological reconstructions for this region are correct, then one might expect a decline in the abundance and diversity of megafaunal grazers after the climatic cooling that culminated in the LGM. However, the fossil vertebrate data do not support this model. Figure 2.21 shows the relative abundance of the bones of woolly mammoth, Pleistocene horse, musk ox, caribou, and Pleistocene bison from a site on the Alaskan North Slope (Matheus, Kunz, and Guthrie 2003). The figure represents radiocarbon-dated bones of the various species. There is a peak in abundance of most taxa at about 35 cal ka, which may be an artifact of unusually good bone preservation during interstadial times.

The subsequent decline in dated remains during the LGM may reflect declining population numbers, or it may be an artifact of poor bone preservation in the cold, dry climates of the glacial period. The study by Matheus, Kunz, and Guthrie (2003) is a cautionary tale, highlighting the potential pitfalls of reconstructing past animal populations based on numbers of preserved bones. Nevertheless, it points out that all these megafaunal mammal species *persisted* in Arctic Alaska through the LGM interval. Following the LGM, it appears that populations of Pleistocene horse, caribou, and bison increased in northern Alaska.

The composition of the late Pleistocene megafauna was somewhat different in western and eastern Beringia (table 2.1). Western Beringia contained twenty-four megafaunal mammal species, including twelve species of grazers (Vereshchagin and Kuz'mina 1984), while eastern Beringia had twenty-one megafaunal species, including five grazers (Anderson 1984). Eastern Beringia might therefore appear to have been relatively depauperate in large grazing mammal species, except that it must be remembered that western Beringia was part of a much larger Asiatic steppe region with faunal connections to regions such as Mongolia and Central Asia. In contrast to this, eastern Beringia was cut off from the rest of North America during much of the late Pleistocene by glacial ice.

Table 2.1
Late Pleistocene Megafaunal Mammal Species of Beringia

Taxon	Eastern Beringia	Western Beringia
Family Edentata		
Megalonyx jeffersoni Dem. (Jefferson's ground sloth)	X	
Family Carnivora		
Canis lupus L. (Gray wolf)	X	X
Cuon alpinus Pall. (Dhole)	X	X
Arctodus simus Cope (Short-faced bear)	X	
Ursus arctos L. (Brown bear)	X	X
Ursus maritimus Phipps (Polar bear)		X
Homotherium serum Cope (Scimitar cat)	X	
Panthera leo atrox Leidy (American Pleistocene lion)	X	
Panthera spelaea Gold. (Cave lion)		X
Family Rodentia		
Castoroides sp. (Giant beaver)	X	
Family Perissodactyla		
Equus spp. (Pleistocene horse)	X	
Equus hemionus Pall. (Asiatic wild ass)		X
Equus lenensis Russ. (Lena horse)		X
Coelodonta antiquitatis Blum. (Woolly rhinoceros)		X
Family Artiodactyla		
Platygonus compressus LeC. (Flat-headed peccary)	X	
Camelops hesternus (Leidy) (Western camel)	X	
Alces alces L. (Moose)	X	X
Cervus elaphus L. (Elk or red deer)	X	X
Rangifer tarandus L. (Caribou or reindeer)	X	X
Bison priscus Boj. (Steppe bison)	X	
Bos baikalensis N. Ver. (Baikal yak)		X
Bos primigenius Boj. (Aurochs)		X
Capreolus pygargus Pall. (Siberian roe deer)		X
Gasella gutturosa Gmel. (Mongolian gazelle)		X
Megaloceros giganteus Blum. (Giant deer, Irish elk)		X
Moschus moschiferus L. (Musk deer)		X
Ovibos moschatus Zimm. (Musk ox)	X	X
Symbos cavifrons Leidy (Woodland musk ox)	X	
Ovis ammon L. (Argall sheep)		X
Ovis dalli Nelson (Dall's sheep)	X	
Ovis nivicola Eschsch. (Snow sheep)		X
Parabubalis capricornis Gromova (Goat-horned antelope)		X
Saiga tatarica L. (Saiga antelope)	X	X
Spirocerus kiakhtensis Pavl. (Twisted-horn antelope)		X
Proboscidea		
Mammuthus primigenius Blum. (Woolly mammoth)	X	X
Mammut americanum (Kerr) (Mastodont)	X	

NOTE: X indicates presence

Recent plant macrofossil studies have shed more light on the nature of Beringian vegetation. Although most paleobotanical studies from the late Pleistocene in Beringia have been based on pollen analysis, Anderson, Bartlein, and Brubaker (1994) observe that the interpretation of ancient tundra vegetation based on fossil pollen spectra is inherently difficult because of low taxonomic resolution, poor dispersal of minor pollen types, and the wide ecological tolerances of genera or species that dominated the pollen rain. Goetcheus and Birks (2001) were able to describe the LGM vegetation of the northern Seward Peninsula on the basis of ancient land-surface macrofossils preserved beneath volcanic ash. Overall, they found that the vegetation was a closed, dry, herb-rich tundra with a continuous moss layer, growing on calcareous soil that was continuously supplied with loess. They interpreted the soils of the Seward Peninsula during the LGM as relatively fertile, being sustained by nutrient renewal from loess deposition and the occurrence of a continuous mat of acrocarpous mosses. There are no exact modern analogues for this vegetation because full-glacial environments and climate with loess deposition do not occur today. It remains to be demonstrated that the vegetation preserved in Goetcheus and Birks's site represents the steppe-tundra vegetation envisioned by paleobotanists as having dominated many regions of Beringia in the late Pleistocene.

What were the large grazers eating? Guthrie (1990, 2001) published the results of studies of plant macrofossils extracted from the molars of late Pleistocene Beringian megafaunal mammals and from the frozen gut contents of woolly mammoths, preserved in permafrost in Siberia. Of the plant remains associated with woolly mammoth, steppe bison, and woolly rhinoceros, more than 75 percent were from grass species. Pleistocene horse molars yielded roughly 40 percent grass remains and 40 percent woody plant remains, with the remainder divided between sedges, forbs, and cryptogams. Fossil musk ox molars yielded 50 percent woody plant remains, about 40 percent grass remains, and about 10 percent sedge remains, which is consistent with their mixed diet today.[21]

These grazers not only needed lots of herbs, they needed nutrient-rich herbs in their diets. Were Pleistocene soils in Beringia more or less rich in nutrients than they are today? Walker et al. (2001) added some modern ecological insights to the debate on nutrient content of ancient steppe-tundra. They suggested that moist nonacidic tundra (MNT) growing today on the Alaskan North Slope may have some important similarities with the ancient Beringian steppe-tundra. MNT grows on calcium-rich, fine-grained soils with relatively high pH. Compared with tussock tundra, MNT soils have ten times the extractable calcium in the active layer, half the organic layer thickness, and 30 percent deeper active layers. Moreover, MNT has twice the vascular-plant species richness, maintains greater habitat diversity, and contains plants with fewer antiherbivory chemicals and more nutrients (particularly calcium). These aspects of the nature of MNT may help us gain a better understanding of the ecology of ancient steppe-tundra.

Boris Yurtsev (2001) also emphasized that Beringia had a high diversity of herbaceous vegetation (grasses, sedges, and forbs) in the mosaic of steppe-tundra landscapes. Based on remnants of steppe vegetation in northeast Asia, Yurtsev described some of the principal types of steppe-tundra vegetation of Beringia as follows: (1) dry watersheds and slopes had cryophytic (cold-adapted) steppes and cryoxerophytic (cold and dry-adapted) herbaceous and prostrate shrub-herbaceous communities; (2) depressions and valleys were occupied by dry steppe-meadows and brackish-water moist meadows; (3) valley meadows and slope pediments were the most productive as pastures for ungulates due to the redistribution of moisture and nutrients within landscapes; and (4) the low-lying

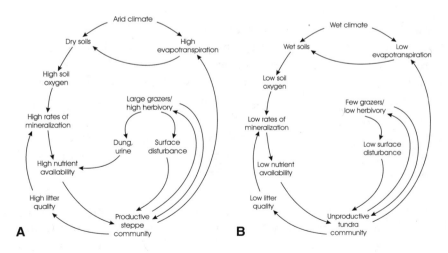

Figure 2.22 Ecosystem models for (A) Beringian steppe-tundra and (B) postglacial mesic tundra environments (after Zimov et al. 1995)

parts of the Bering Land Bridge were covered with shrub tundra, which served as a barrier for the dispersal of steppe plants and animals.

If steppe-tundra dominated the interior regions of both western and eastern Beringia during much of the late Pleistocene, how did it maintain itself, and what caused it to be replaced by modern arctic tundra at the end of the last glaciation? Two intriguing theories about these phenomena have been published—one by Zimov et al. (1995) and the other by Guthrie (2001).

The Zimov et al. (1995) model suggests that the presence of large numbers of megafaunal grazers was at least partially responsible for the maintenance of steppe-tundra communities (figure 2.22). In this model, arid climatic conditions produce high evapotranspiration rates that foster soil aridity (figure 2.22 A). Dry soils have relatively high oxygen levels, which lead to high rates of mineralization. Highly mineralized soils have high nutrient availability, which supports productive steppe vegetation. The herds of megafaunal grazers that feed on the grasses trample the surface repeatedly, and this trampling and grazing favors persistence of grasses on the landscape, as opposed to the moss-dominated ground cover of modern arctic tundra. In the Zimov et al. model, human predation and increased effective moisture in Beringia at the end of the Pleistocene tipped the balance toward modern arctic tundra communities. In these communities, low evapotranspiration leads to wet soils that hold little oxygen (figure 2.22 B). Low rates of soil mineralization lead to low nutrient availability and a relatively unproductive tundra vegetation, supporting few grazers and low levels of herbivory.

Guthrie (2001) emphasizes the importance of clear skies over Beringia during the late Pleistocene. This would have enhanced evapotranspiration from the surface during summer days and caused a radiation deficit to clear night skies in winter, leading to lower temperatures. During times of low insolation caused by orbital forcing (e.g., the LGM), the west Beringian steppe-tundra (as part of the larger Asiatic steppe) was kept arid by a huge, stable, high-pressure system north of the Tibetan Plateau. North Atlantic sea ice cover and the Scandinavian Ice Sheet would have cut off the main source of westerly moisture, enhancing aridity in Asiatic steppe regions. Lowered sea levels and the Cordilleran Ice Sheet would have reduced the moisture available to interior regions of eastern Beringia.

The high-pH soils described by Höfle and Ping (1996) from the ash-buried surfaces on the Seward Peninsula are considered by Guthrie to be typical of steppe-tundra environments. As discussed by Walker et al. (2001), these soils were probably deeply thawed in summer, were rich in nutrients, and contributed to higher primary productivity. Guthrie attributes the late glacial shift from steppe-tundra to modern arctic tundra to a change in regional climates. Decreasing aridity was brought about by the melting of ice sheets and sea ice, the inundation of the Bering Land Bridge as sea level rose, and most of all by the northerly penetration of dense, low-lying cloud cover, which greatly reduces net insolation at the surface level in the high latitudes. Finally, commenting on the Zimov et al. (1995) model, Guthrie acknowledges that grazers can influence vegetation, but he questions whether they can do so at such a high level over such a large region.

LGM ENVIRONMENTS OF BERINGIA

During the Last Glacial Maximum (LGM), most of Beringia experienced very cold, arid climate. In western Beringia, glacial ice covered high mountain regions and cirques and spilled out into extensive valley glaciers in the Priokhot'e region and southeastern Chukotka, as well as in the headwater regions of the Indigirka and Kolyma Rivers (Brigham-Grette et al. 2004). However, as shown in figure 2.23, most of the lowland regions of western Beringia remained essentially ice-free during the last glaciation. A similar situation occurred in eastern Beringia, where glaciers grew in regional mountain ranges and only succeeded in covering lowlands south of the Alaska Range. This southern Alaskan ice sheet formed the northwestern limits of the Cordilleran Ice Sheet that blanketed much of western Canada. The lowland regions remained ice-free because aridity deprived mountain glaciers of the moisture necessary to expand their margins significantly.

Paleotemperature estimates based on fossil beetle assemblages (Elias 2001a) (figure 2.24; table 2.2) suggest that average summer temperatures (TMAX) were as much as 7.5°C cooler at 22 cal ka than they are today at Bluefish Caves in the northern Yukon. However, it appears that this cooling of summer temperatures was far from uniform across Beringia. On the Seward Peninsula, for instance, TMAX was depressed by a maximum of 2.9°C, and the beetle temperature estimate

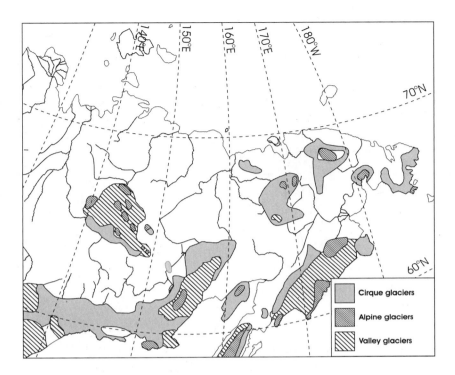

Figure 2.23 Glaciated regions of western Beringia during the Last Glacial Maximum (after Brigham-Grette et al. 2004)

ranges upward into the positive zone, so that TMAX may have been 1.1°C *warmer* at 20 cal ka than it is today at that site. Most eastern Beringian paleotemperature estimates for the LGM, based on beetle data, fall within a range of 3–5°C colder than modern TMAX. This is somewhat surprising, because there is ample evidence from the mid-latitudes of the Northern Hemisphere that TMAX was depressed by 10–11°C during the LGM, based on beetle evidence from North America and Europe (Elias, Anderson, and Andrews 1996; Coope et al. 1998). Summer temperature depressions of 10°C have also been interpreted from the oxygen isotope record of Greenland ice cores (Severinghaus and Brook 1999). Mean January temperatures (TMIN) in eastern Beringia were probably depressed by very little, if at all (Elias 2001a) (figure 2.24). The best-constrained TMIN estimates from Alaska and the Yukon were within a few degrees centigrade of modern values.

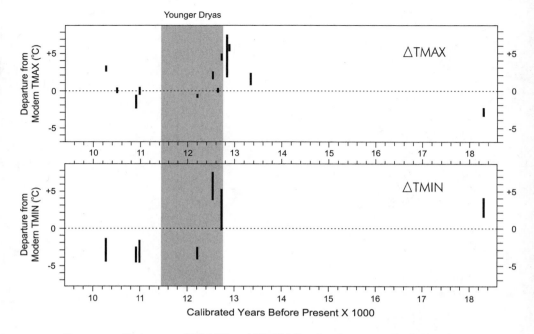

Figure 2.24 Estimates of TMAX and TMIN for sites in eastern Beringia, shown as departures from modern TMAX and TMIN, respectively; note the decline in TMAX during the Younger Dryas interval (data from Elias 2001a)

Alfimov and Berman (2001) studied fossil beetle assemblages from the northeastern part of western Beringia and found that the late Pleistocene assemblages of the Kolyma and Indigirka River Basins could not have survived TMAX levels less than 9–10°C and that TMAX elsewhere in western Beringia was as high as 12–13°C. In other words, average summer temperatures in western Beringia were essentially as high as they are today in many regions. They argued for greater continentality of regional climates throughout the Late Pleistocene in their study region.

Sher et al. (2002) added some refinements to Alfimov and Berman's work. With data from additional sites, including the exposed Laptev Sea shelf, Sher et al. performed a new reconstruction of TMAX and TMIN for northeastern Siberia, based on fossil insect assemblages. In their estimation, TMAX remained at or above modern levels even during the LGM and TMIN values remained close to modern levels as well. However, they noted that other climate proxies indicate a 10°C cooling of TMIN during the LGM in western Beringia and that insects are not reliable proxies for winter temperatures because they are not active in

Table 2.2

Mutual Climatic Range (MCR) Estimates and Departures from Modern Maximum and Minimum Temperature Values (°C; TMAX and TMIN) for Late Pleistocene Insect Fossil Assemblages in Eastern Beringia

Site and Sample	Age (Years BP)	July			January			Reference(s)
		TMAX	TMAX Modern	ΔTMAX	TMIN	TMIN Modern	ΔTMIN	
Rock River, Yukon	25,400	10.5–11.5	16.0	−5.5 to −4.5	−28 to −22.5	−29.0	+1 to +6.5	Matthews and Telka 1997
Eva Creek 3-3C, Alaska	24,600	11–13.25	16.9	−5.9 to −3.65	−23 to −17.5	−23.4	+0.4 to +5.9	Matthews 1968
Bering Shelf 78-15	22,909 ± 703	5.5–10.5	10.8	−5.3 to −0.3	−34.5 to −18.75	−15.6	−18.9 to −3.15	Elias, Anderson, and Andrews 1996; Elias, Short, and Birks 1997
Bluefish Caves, Yukon	22,073 ± 593	8.5–9.5	16	−7.5 to −6.5	−31.5 to −28	−29.0	−2.5 to +1	Matthews and Telka 1997
Bering Land Bridge Park, Alaska	19,863 ± 330	5.5–9.5	8.4	−2.9 to +1.1	−34 to −21	−20.3	−13.7 to −0.7	Elias 2001a
Colorado Creek, Alaska	16,307 ± 404	11.25–12.5	14.8	−3.55 to −2.3	−21.25 to −18.25	−22.6	−1.35 to +4.35	Elias 1992

winter months in cold regions. However, as with other research topics in Beringia, not all researchers agree on these temperature reconstructions. For instance, other lines of evidence such as periglacial features that developed during the LGM indicate that mean annual temperatures dropped significantly (Hopkins 1982). Most paleoclimatologists would agree that, generally speaking, Beringia was relatively dry and cold with cooler summers during the LGM. More meso-scale patterns indicate east-to-west trends in temperature and moisture gradients, with colder and drier conditions dominant over eastern Beringia (Anderson and Brubaker 1994; Lozhkin et al. 1993; Anderson et al. 1997; Carter 1981; Hamilton 1994; Brigham-Grette et al. 2004). The discrepancies among geomorphological, palynological, and entomological reconstructions indicate that further research is needed to clarify LGM environments in Beringia (Elias 2001b)

What vegetation cover developed in Beringia during the LGM? As discussed in this chapter, there is little controversy concerning the vegetation cover of western Beringia: steppe-tundra dominated most lowland regions. But there is no consensus about LGM vegetation cover in eastern Beringia. Most paleobotanical studies from this interval have focused on pollen extracted from cores. As Anderson, Bartlein, and Brubaker (1994) point out, the interpretation of ancient tundra vegetation based on fossil pollen spectra is inherently difficult because of low taxonomic resolution, poor dispersal of minor pollen types, and the wide ecological tolerances of genera or species that dominated the pollen rain. However, Goetcheus and Birks's (2001) study of the buried surface LGM vegetation from the Seward Peninsula yielded some useful insights. Fossil studies from southwestern Alaska and certain localities on the Bering Shelf (the former Bering Land Bridge surface) give indications that mesic shrub tundra persisted in parts of central Beringia, even during the LGM.

Schweger's (1997) review of late Pleistocene environments in the Yukon argues for regional LGM vegetation with affinities to two modern ecosystems: herb-dominated tundra found today in the North American Arctic, and Eurasian steppe. He also argues that the Yukon experienced a harsher, more arid climate than did western Alaska and the Bering Land Bridge. Summer temperatures were cooler than they are now, and winter temperatures were warmer than today. Snow cover was thin and discontinuous. This paleoclimatic reconstruction agrees well with the fossil insect–based reconstructions of summer and winter temperatures in Alaska.

Given that shrub tundra was at least regionally important in Beringia, what can we say about the nature of the vegetation? Until we have a fuller understanding of ancient shrub tundra communities, the best approach may be to examine the modern analogs. Dwarf birch and willow are important elements of modern shrub tundra in Alaska, the Yukon, and northeastern Siberia. An analysis of their role in such modern tundra plant communities sheds light on their possible role in ancient Beringian ecosystems. Based on modern plant ecological studies from throughout the arctic regions of the Northern Hemisphere (CAVM Team 2003), various kinds of shrub tundra communities dominate many regions. Dwarf willow and birch are found in a wide variety of these plant communities. For instance, *Salix glauca* and *S. callicarpaea* are both found in erect dwarf-shrub tundra communities, characterized by dry to moist acidic soils. Dwarf birch also grows in this community, including *Betula nana/exilis* and *B. glandulosa*. In addition, dwarf willow and birch also grow in low-shrub tundra. In this community type, willows dominate wet habitats (including *Salix pulchra, S. glauca, S. richardsonii, S. alaxensis, S. krylovii, S. burjatica, S. boganidensis,* and *S. arbusculoides*) associated with peatlands and permafrost, while the dwarf birch (including *Betula nana/exilis,* and *B. middendorfii/glandulosa*) grows on the drier uplands.

In a third dwarf shrub tundra community, called graminoid, prostrate dwarf shrub–forb tundra, the willows include *Salix polaris, S. rotundifolia, S. arctica,* and *S. reticulata.* This is moist to dry tundra on fine-grained, nonacidic soils with moderate snow. Plant cover is moderate (40–80 percent) in this community. Even on dry tundra, willows play an important part. For instance, on prostrate dwarf shrub–herb tundra, the dominant shrubs are *Salix arctica, S. polaris, S. rotundifolia,* and *S. phlebophylla.* This plant community can be quite patchy, with as little as 20 percent plant cover. Lichens and mosses cover from 30 to 60 percent of the landscape. This brief survey of modern arctic shrub tundra plant communities demonstrates the ecological amplitude of dwarf willow and birch as groups of species.

Within this assortment of dwarf shrubs, one can find species adapted to almost all conceivable landscape types of late Pleistocene Beringia, but especially the moister patches of the landscape. As Yurtsev (2001) asserted, the areal coverage of these mesic and damp habitats may have been diminished during the arid, full-glacial intervals in Beringia, but they were never fully extinguished. As discussed next, there is good fossil evidence that the various components of shrub tundra persisted throughout Beringia, even during the LGM, and flourished during the subsequent Lateglacial interval.

SYNTHESIS OF LATEGLACIAL ENVIRONMENTS

The Lateglacial period (17–12.8 cal ka) was an interval of rapid environmental change throughout Beringia and, indeed, throughout much of the world. Climatic fluctuations brought about wholesale changes in the distribution of Beringian plants and animals and may have had the most important role in the regional extinction of many megafaunal mammal species.

According to the fossil beetle data from eastern Beringia (Elias 2001a), TMAX values began rising by about 13.8 cal ka, reaching warmer-than-modern levels by 12.8 cal ka. The pollen evidence (Brubaker, Anderson, and Hu 2001; Edwards et al. 2001; Bigelow and Edwards 2001) indicates that herbaceous tundra vegetation dominated much of Beringia at the end of the last glaciation, giving way to shrub tundra in most regions between about 19 and 13.5 cal ka.

This transition was not synchronous throughout Beringia, however. As shown in figure 2.25, the earliest recorded transition from steppe-tundra to more mesic tundra vegetation occurred in western Beringia, at Patricia Lake (Anderson and Lozhkin 2002) at 19.3 cal ka. This is well before the generally accepted time of climatic amelioration that marked the transition from LGM to Lateglacial climates. A recent study of radiocarbon-dated pollen assemblages from sites throughout Beringia (Brubaker et al. 2005) concluded that most, if not all, of the woody plant species dominant today in Alaska, the Yukon, and northeastern Siberia were present in Beringia during the LGM, albeit in much smaller numbers. The study focused on *Populus, Larix, Picea, Pinus, Betula,* and *Alnus/Duschekia.* Unfortunately, it did not consider the history of *Salix,* because the tundra (shrub) and boreal (tree) species cannot be distinguished on the basis of their pollen grains. One of the difficulties in Beringian paleobotany is that

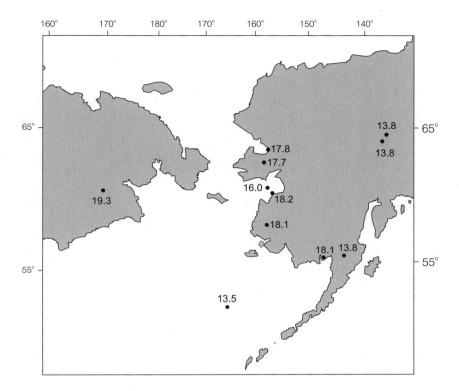

Figure 2.25 Earliest published dates (expressed in calibrated years BP × 1,000) for the transition from steppe-tundra to more mesic shrub tundra vegetation in Beringia (data sources are cited in the text)

several of the important plant species exhibit multiple growth forms. For instance, *Betula glandulosa* can be a prostrate shrub, less than 50 centimeters tall, or it can be an upright, multistemmed woody plant that reaches almost 2 meters in height. It also hybridizes with tree birch (*B. neoalaskana*) to form trees several meters tall (Edwards, Dawe, and Armbruster 1991). *Salix alaskensis* varies from a prostrate shrub in the arctic tundra to a 6–9-meter tall tree in the boreal forest (Viereck and Little 2007).

The general trend in western Beringia was that by 14.7 cal ka, steppe-tundra gave way to new vegetation. In the southern regions, this new vegetation was larch forest. Larch does not exhibit a prostrate shrub growth form. Siberian larch trees range between 15 and 50 meters in height. In the northern regions, the new vegetation was high-shrub tundra (Lozhkin et al. 1993). The pollen evidence suggests that shrub tundra gave way to more forested conditions by about 11.5 cal ka (Anderson and Lozhkin 2002). Based on plant macrofossil records, the latter vegetation type included tree-sized willows and alders (*Duschekia*) (Edwards et al. 2005).

On the Bering Land Bridge, the transition from herbaceous to mesic tundra was registered in the pollen record at a site in Norton Sound (Ager 2003) at about 16 cal ka, but close by on St Michael Island at 18.2 cal ka. Dates in excess of 18 cal ka are associated with the transition throughout southwestern Alaska, and in the Seward Peninsula the transition apparently took place a few centuries after 18 cal ka (Ager 2003).

In interior Alaska and at the base of the Alaskan Peninsula, the transition from herb-zone pollen spectra to shrub or mesic tundra took place somewhat later, about 13.8 cal ka (Bigelow and Edwards 2001; Brubaker, Anderson, and Hu 2001; Edwards et al. 2001). A few centuries later, this transition to more mesic tundra also occurred in the Pribilof Islands region, which was then situated near the outer margin of the land bridge.

Climatic perturbation was a major driving force on the Beringian landscapes of the late glacial interval. Toward the end of the Lateglacial interval, there is evidence of a large-scale climatic oscillation in some parts of Beringia. The story is far from complete, but we catch glimpses of the pace and intensity of environmental change from specific localities. For instance, in the Mackenzie Mountain region, which lay at the easternmost edge of Beringia, Szeicz and MacDonald (2001) found evidence of rapid climatic amelioration by 12.9 cal ka. *Populus* expanded in these mountains from about 12.9 to 10.1 cal ka. During this same interval, *Populus* expanded into north-central Alaska (Anderson and Brubaker 1994). In the early Holocene (ca. 10.7 cal ka), *Populus* expanded onto the Alaskan North Slope (Nelson and Carter 1987). The latter expansion was inferred to have happened under a warmer-than-modern thermal regime in which mean summer temperatures were 2–3°C higher than today. Brubaker et al. (2005) consider that, from 14 to 11 cal ka, *Populus* (mostly likely balsam poplar) expanded from restricted populations in Beringia. Its first appearance in the fossil pollen record was at 16 cal ka in the Brooks Range. Macrofossil evidence indicates that trees of balsam poplar and aspen (*Populus tremuloides*) were widespread in eastern Beringia by 11.5 cal ka (Edwards et al. 2005).

In western Beringia, a northward expansion of high-growth forms of willow took place by about 13.5 cal ka (Lozhkin et al. 1993). At that time, large willows were growing at least 100 kilometers further north than they do today, suggesting an interval of warmer-than-modern climate. However, the most recent evaluation of the western Beringian paleobotanical data (Edwards et al. 2005) has concluded that shrub tundra dominated western Beringia during the Lateglacial interval until about 11.5 cal ka.

Macrofossil evidence indicates that at least the low-shrub, prostrate form of willow persisted throughout much of Beringia, even during the LGM. Willows are both wind- and insect-pollinated, even within single species (Peeters and Totland 1999). Their pollen production and dispersal are moderate, so they tend to be somewhat underrepresented in fossil pollen assemblages. Dwarf willow and dwarf birch often grow together in mesic shrub tundra communities, and in some ways the presence of dwarf birch in pollen assemblages can be taken as a crude proxy for dwarf willow. Because birch is overrepresented in the pollen rain of arctic regions, the interpretation of ancient birch localities has its own set of problems.

The pollen mapping work of Brubaker et al. (2005) indicates that birch also survived in situ in Beringia during MIS 2. The growth form of birch cannot be determined from its pollen in Beringia, but the most likely candidate for regional LGM birch was the shrub form (dwarf birch) and the earliest birch macrofossils following the LGM are from shrub birch (Edwards et al. 2005). The fossil pollen data indicate that shrub birch had a patchy distribution in Beringia during the last glaciation. This distribution was probably controlled by a combination of environmental factors, including soil conditions, hydrology, and microclimate. During 16–14 cal ka, shrub birch began to spread into many regions. This spread has been attributed to increasing temperatures and precipitation (Brubaker et al. 2005).

Alder (*Alnus/Duschekia*) probably also survived in situ in Beringia during the LGM (Brubaker et al. 2005), although its pollen record is quite difficult to interpret. Alder shrubs are part of the tall shrub vegetation type in the arctic today. These generally grow to heights greater than 2 meters. Their expansion began by about 14 cal ka in western Beringia, but not until 9–8 cal ka in eastern Beringia (Brubaker et al. 2005).

Coniferous forest expansion in Beringia came at different times in different regions. In eastern Beringia, conifer forests started to expand only in the early Holocene. This may have been because the only populations of conifers available to colonize Alaska were growing in distant regions of the Yukon during the last glaciation. However, Brubaker et al. (2005) argue against this scenario. Their interpretation of the mapped pollen data for Beringia indicates that both spruce and pine survived the LGM in situ in Beringia. White spruce was the most likely species seen in LGM pollen spectra from eastern Beringia, but black spruce accompanied white spruce in becoming established at some sites in northern British Columbia (Pisaric et al. 2001). Orbital parameters in northern latitudes were yielding extremes of summer insolation and minimal winter insolation at 12.9 cal ka, so the relatively high degree of climatic continentality may have had a role in the slow migration of conifers. Paleohydrological modeling by Edwards et al. (2001) suggests that the Lateglacial period in eastern interior Alaska was a time of relative aridity, so the combination of extremely cold winters and little effective moisture may have been important elements in limiting the expansion of coniferous forests in eastern Beringia.

In western Beringia, Siberian pine (*Pinus pumila*) shrubs probably grew in scattered localities during the LGM, according to the pollen data (Brubaker et al. 2005). Larch apparently survived the LGM in the southern regions. Plant macrofossils and pollen from the Yana and Indigirka Lowlands date as early as 16 cal ka, possibly indicating a second, northern refugium for larch in western Beringia (Lozhkin et al. 2002). Open larch (*Larix dahurica*) forest began to spread before 13.4 cal ka (Lozhkin et al. 1993). Based on the multiple refugium theory of Brubaker et al. (2005), this spread may have come through multiple source areas in western Beringia. Whatever the source, apparently this spread of larch forest was uninterrupted by the kind of climatic oscillation (Younger Dryas cooling) evidenced in eastern Beringia.

THE YOUNGER DRYAS INTERVAL

A growing body of evidence points toward a climatic cooling during the Younger Dryas chronozone in eastern Beringia (12.8–11.4 cal ka) (Brubaker et al. 2001). Elias (2001a) found evidence of a dramatic decline in TMAX values during this interval, especially in arctic beetle assemblages. Bigelow and Edwards (2001) noted a reduction in shrub tundra and corresponding increase in herb tundra in central Alaska between 12.4 and 11.2 cal ka. Climatic fluctuations during the Younger Dryas also have been reported from southern Alaska (Engstrom, Hansen, and Wright 1990; Hansen and Engstrom 1996; Hu, Brubaker, and Anderson 1995). The fluctuations have been interpreted as cooling and drying. The changes in pollen spectra at some of these sites were subtle and difficult to interpret. More substantial evidence for Younger Dryas cooling has been reported from the pollen record for Kodiak Island (Peteet and Mann 1994).

The reconstruction of regional events in the Younger Dryas interval is fraught with difficulties. First among these is the difficulty in dating events because the interval from 12.8 to 11.4 cal ka falls in a "radiocarbon age plateau." Because of changes in the rate of atmospheric production of ^{14}C during this interval, and the release of ancient carbon that had been frozen in ice and sediments during the LGM, radiocarbon ages from the Younger Dryas interval are unreliable. Independent calibrations of the radiocarbon curve have shown that ^{14}C ages plateau at 12.5 cal ka and again at 11.8–11.7 cal ka. The latter plateau persisted for about 250 calendar years (Hughen et al. 2000).

While the relative strength of climatic cooling during the Younger Dryas interval may have varied considerably throughout eastern Beringia, there is very little evidence of such a cooling in western Beringia. Lozhkin et al. (2001) noted that evidence for a Younger Dryas–type climatic event is absent from Wrangel Island and from most northern and eastern regions of western Beringia. However, Pisaric et al. (2001) reported a possible Younger Dryas cooling in the lower Lena River Basin, based on pollen evidence indicating an increase in herbaceous tundra at the expense of shrub tundra.

If there was a major reexpansion of steppe-tundra in eastern Beringia during the Younger Dryas interval, what was its cause? It would seem that climatic cooling per se would not have been sufficient to cause this type of change in vegetation. Perhaps the combination of rapid cooling *and* decreased precipitation was responsible. We suggest that one mechanism for bringing about a decrease in precipitation throughout much of eastern Beringia during the Younger Dryas interval may have been increased sea-ice cover in the North Pacific. Expanded sea-ice cover would have blocked much of the flow of Pacific

Table 2.3
Late Pleistocene Extinctions of Megafaunal Genera in North America

Taxon	Last 100	100–50	50–16	16–11.5	<11.5
Glyptodont					
Glyptotherium	X				
Ground sloths					
Megalonyx				X	
Eremotherium	X				
Nothrotheriops				X	
Glossotherium			X		
Giant armadillo					
Pampatherium				X	
Giant beaver					
Castoroides				X	
Capybaras					
Hydrochaeris	X				
Neochoerus	X				
Bears					
Arctodus				X	
Tremarctos				X	
Cats					
Homotherium			X		
Miracinonyx				X	
Panthera				X	
Smilodon				X	
Mammoths and mastodonts					
Mammuthus				X[b]	X[c]
Cuvieronius			X		
Mammut				X	
Horses					
Equus				X	
Tapir					
Tapirus				X	
Peccaries					
Mylohyus				X	
Platygonus				X	
Camels					
Camelops				X	
Hemiauchenia				X	
Paleolama				X	

Timing of Extinction (cal ka)[a]

(Table 2.3 continued)

Taxon	Last 100	100–50	50–16	16–11.5	<11.5
			Timing of Extinction (cal ka)[a]		
Deer					
Bretzia	X				
Cervalces				X	
Navahoceros	X				
Torontoceros	X				
Pronghorn antelope					
Stockoceros	X				
Tetrameryx	X				
Musk ox, Shrub ox, Saiga antelope (Bovidae)					
Bootherium	X				
Euceratherium				X	
Saiga	X[b]			X[d]	

SOURCE: After Barnowsky et al. 2004

[a] X indicates extinction; placement in age columns is based on the youngest reliably dated occurrence
[b] Time of extinction in the lower forty-eight states
[c] A relict population of woolly mammoth remained extant until 8.8 cal ka on St. Paul Island, Alaska (Guthrie 2004)
[d] Time of extinction of saiga antelope in eastern Beringia

moisture into interior regions. This hypothesis remains to be tested by paleooceanographic research, which has been surprisingly scarce in our study region.

PLEISTOCENE EXTINCTIONS

The effects of rapid environmental change on the megafauna of Beringia were quite dramatic. The youngest dates reported by Matheus, Kunz, and Guthrie (2003) for megafaunal species on the Alaskan North Slope are as follows: mammoth, 14.7 cal ka; horse, 14.6 cal ka; and bison, 12.3 cal ka. Both woolly mammoth and Pleistocene horses appear to have died out in eastern Beringia as the steppe-tundra habitat gave way to mesic shrub tundra. At the end of the Lateglacial interval, bison, moose, caribou, and musk ox would have been the only prey animals left for big-game hunters in eastern Beringia. Bison persisted in Alaska and the Yukon until the last century, but their range shifted south of the arctic in Alaska and the Yukon during the early Holocene (Stephenson et al. 2001). Radiocarbon-dated bison bones younger than 12.3 cal ka are all from sites south of the Brooks Range.

A major debate in Quaternary paleoecology and archaeology concerns the role of human predation in the extinction of megafaunal mammals at the end of the Pleistocene. As discussed by Martin (1989), humans may have played a major part in the extinction of dozens of species of megafaunal mammals, particularly on the continents where humans were late arrivals on the landscape and their potential prey animals had no instinctive fear of them. At the opposite end of this debate, researchers such as Grayson and Meltzer (2003) argue that human hunters had virtually no role in the extinction of the Pleistocene megafauna. The debate is far from settled, and much of the difficulty in reaching a consensus lies in the conundrum of building a reliable chronology of the two sets of phenomena: the arrival of humans in North America, and the timing of extinction of the various species of megafauna.

Two basic models are proposed by those who believe that human hunters had an important role in the demise of the megafauna. In the "Overkill" model, human hunting precipitated extinction by causing the death rate of the hunted species to exceed the birth rate. This can be a relatively slow process, lasting 1,500 years or more. The "Blitzkrieg" model is a special case of overkill, in which human hunting pressure on megafaunal populations was so intense that it caused extinction in just a few centuries. This type of extinction would have occurred more rapidly along a geographic front where humans came into contact with prey animals.

Brook and Bowman (2002) developed a mathematical model to simulate the relative importance of prey naivety in the effectiveness of human predation. They argue that previous models, such as one by Alroy (2001), assumed complete and unchanging prey naivety throughout the interval when the extinctions were taking place. While Brook and Bowman reject the idea that the Blitzkrieg model has been proved, they do conclude that

> inferring robustly the cause of extinction of the Pleistocene megafauna is a remarkably complicated problem that is very sensitive to assumptions concerning the analysis and interpretation of existing data. Although great progress has been made, it is premature to suggest that the problem has been cracked. (Brook and Bowman 2002:14,627)

The most recent attempt to solve this problem is by Barnowsky et al. (2004), who evaluated data from many disparate sources, including paleontology, climatology, archaeology, and paleoecology. They noted that twenty-eight out of forty-one genera of megafaunal mammals (68 percent) became extinct in North America at the end of the Pleistocene (table 2.3). Of these, five genera died out in North America but managed to survive on other continents (e.g., saiga, which became extinct in eastern Beringia but survives in central Asia today). In mid-latitude North America, the megafaunal extinction event ties closely with the Clovis culture, the first well-documented group of people that lived south of the continental ice sheets. This so-called Clovis window falls between 13.3 and 12.8 cal ka.

Barnowsky et al. (2004) noted that in Alaska megafaunal extinction had taken place "even in the absence of significant human populations" and was therefore more likely to have been caused by environmental change. For instance, in Alaska and the Yukon, certain species of horses and the short-faced bear became extinct about 36.5 and 25.3 cal ka, respectively. Datable mammoth fossils dropped in abundance across Europe and Alaska after 13.8 cal ka, with the last dated remains from interior Alaska at 13 cal ka. Arguably, this indicates a decrease in effective population sizes then; however, mammoths survived until 11 cal ka on the Taimyr Peninsula, until 4 cal ka on Wrangel Island, and until 8.8 cal ka on St. Paul Island in the Bering Sea (Guthrie 2004). Data from Guthrie (2006) indicate that Pleistocene horse died out in eastern Beringia by 14.5 cal ka. Pleistocene bison persisted in eastern Beringia until 10 cal ka (figure 2.20).

Ultimately, Barnowsky et al. (2004) concluded that humans contributed to megafaunal extinctions in North America but that human hunting was not solely responsible. They also noted that firmer chronologies, more realistic ecological models, and regional paleoecological insights are still needed to understand the details of megafaunal extinction patterns.

EARLY HOLOCENE ENVIRONMENTS

The early Holocene was a time of continued climatic amelioration and massive environmental change. When the Bering Land Bridge flooded with seawater (after 12.9 cal ka), Beringia ceased to exist as a coherent biological region.

Although larch forests had spread through much of western Beringia before the beginning of the Holocene, in the lower Lena Basin of Siberia, spruce and larch expanded after about 9.5 cal ka (Pisaric et al. 2001). On the easternmost edge of Beringia, in the Mackenzie Mountains of Canada, spruce expanded upslope by 8.9 cal ka (Szeicz and MacDonald 2001). Spruce forests were present in central Alaska at the same time (Bigelow and Edwards 2001). Modeling suggests that, during this interval, climates of interior Alaska were warmer and drier than modern climates (Edwards et al. 2001). The spread of coniferous forests across Alaska was reasonably rapid in some regions and incredibly slow in others. Recent models of tree dispersal suggest that high rates of spread are not feasible (J. S. Clark et al. 2003), which is borne out by the fossil record from parts of eastern Beringia. For example, spruce forest did not reach the end of its migration route in southwestern Alaska until as late as 5.2 cal ka (Brubaker, Anderson, and Hu 2001).

Settlement of Northern Asia

The settlement of Beringia can be understood only within the wider context of the human colonization of northern Asia. Although climate change at the end of the Pleistocene and its effect on biota was a factor, the movement of people into areas east of the Verkhoyansk Mountains—and across the now-submerged continental shelf area between Chukotka and western Alaska—was primarily a consequence of the evolved ability of humans to occupy environments outside the tropical zone.

The process began with human expansion into mid-latitude regions of North Africa and Eurasia after 1.8 million years ago by early representatives of *Homo*. This was followed by initial occupation of areas above latitude 45°N in western Europe and subsequent expansion of the Neandertals eastward into cooler and drier regions of northern Eurasia. But the most critical development was the evolution of modern humans—again derived from the tropical zone—who dispersed into higher latitudes 50,000–40,000 years ago with a vastly improved capacity for coping with cold climates (Gamble 1994; Hoffecker 2005a).

The current distribution of winter temperatures in the Northern Hemisphere reveals a pattern that was heavily influenced by the presence of land masses and mountain ranges. This is especially evident in Eurasia, where the westerlies bring increasingly cold and dry air to the eastern portion of the continent as they travel away from the North Atlantic, creating both a longitudinal and a latitudinal winter temperature gradient (figure 3.1).[1] The coldest winter air is concentrated in a pocket east of the Verkhoyansk Mountains, where mean January temperatures are −45°C (sometimes described as the "cold pole") (Lydolph 1977:442; Henderson-Sellers and Robinson 1986). The interior regions of Northeast Asia—and particularly the interior of western Beringia—presented a major climate barrier to human settlement, even during the warmest intervals of the past million years.

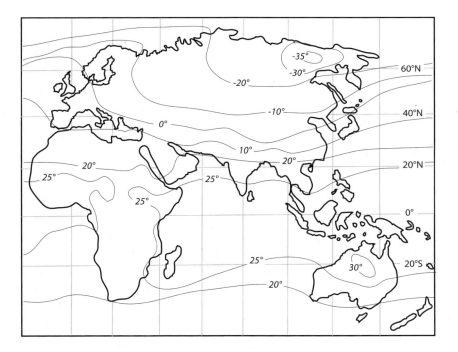

Figure 3.1 Mean winter temperatures in Eurasia, Africa, and Australia, showing the position of the "cold pole" east of the Verkhoyansk Mountains in Northeast Asia (or western Beringia) (based on Henderson-Sellers and Robinson 1986:84, fig. 2.32)

Low winter temperatures, however, are not the only barrier to human settlement in northern Asia. Cold and aridity, along with reduced solar energy (due to high latitude), combine to limit biological productivity in this part of Eurasia. In Northeast Asia, rainfall averages less than 25 centimeters per year. And even in southernmost Siberia (i.e., 50°N latitude), absorbed solar radiation is roughly half the amount received at the equator, whereas energy loss due to albedo increases substantially above this latitude (Young 1994:7–9). Plant productivity is generally low in northern Asia, and it is especially low above 60°N latitude. Mean annual primary productivity (net) for boreal forest is 500 grams per square meter (g/m^2) and for tundra is 140 g/m^2 (compared with a mean annual production of 2,000 g/m^2 for tropical forest) (Leith 1975; Whittaker 1975:192–235; Archibold 1995).[2]

Extreme seasonal variations in northern Asia create further stress for human populations. These variations are a product of both the continental climate regime and high latitude. At Verkhoyansk in the Yana River Basin, which lies at 67°N and averages 15 centimeters of precipitation each year, the difference between the mean July and January temperatures is almost 60°C (i.e., more than 100°F). And winter sunlight is limited across much of northern Asia, disappearing altogether for some period above the Arctic Circle (66° 33' N).

Early Humans in Northern Asia

Humans (or the subfamily Homininae) diverged from the African ape lineage more than 5 million years ago, but for most of their history on earth they have been confined to tropical habitat in Africa below latitude 16°N (Klein 1999:186–187). In terms of climatic range, australopithecines were comparable to the living African apes, who inhabit tropical forest and woodland where the mean annual temperature does not vary significantly from 25°C (Tuttle 1986). It is only after the appearance of representatives of the genus *Homo* (which occurred at 2.5 mya [million years ago]) that humans moved into the middle latitudes. Earliest *Homo* exhibits a significant increase in brain volume and is associated with the oldest identified stone-tool industry (Oldowan) and evidence for meat consumption.

By 1.8 mya, *Homo* sites are found as far as 35°N in North Africa (Aïn Hanech) and 41°N in the southern Caucasus region (Dmanisi) (Sahnouni and de Heinzelin 1998; Gabunia et al. 2000). Occupation of mid-latitude regions in East Asia is currently dated at 1.66 mya for the Nihewan Basin (also at 41°N) in China (Zhu et al. 2001). Major geographic expansion is coupled with evidence for the occupation of new habitats in Africa, including the Ethiopian Plateau (2,300 meters above sea level [asl]), where sites are dated to 1.5 mya (Cachel and Harris 1998) (figure 3.2).

Although skeletal remains from Dmanisi exhibit similarities to earliest *Homo* (Vekua et al. 2002), the movement out of Africa and into the middle latitudes is broadly associated with the appearance of *H. ergaster, H. erectus,* and other later forms (Dennell and Roebroeks 2005). This would link the expansion to changes in postcranial anatomy—such as fully modern bipedal locomotion—that may reflect foraging adaptations to less-productive habitat (Walker and Shipman 1996). *H. erectus* had evolved an energy-efficient means for foraging over large areas where food resources were more dispersed. Also, evidence from

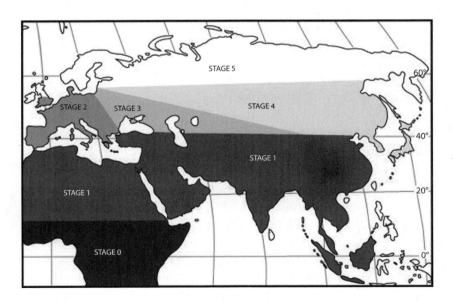

Figure 3.2 Broad patterns of settlement in the middle and higher latitudes during the Pleistocene (between 1.8 mya and 11 cal ka) (from Hoffecker 2005a:7, fig. 1.2)

East Africa suggests increased consumption of meat (Monahan 1996), which represents another adaptation to reduced biological productivity and lower resource density.

Stone tools associated with the initial *Homo* expansion into middle latitudes are confined to pebble choppers, flakes, and other typical Oldowan forms, however. Handaxes and other Acheulean tools do not appear outside Africa until 1.4 mya and never became common in East Asia (Bar-Yosef and Belfer-Cohen 2001). There is no evidence at present that the colonization of new regions during 1.8–1.6 mya was achieved with novel developments in technology (Dennell and Roebroeks 2005).

Early humans occupied higher latitudes in Europe after 0.5 mya (as far as 52°N by 0.3 mya) (figure 3.2: stage 2),[3] but it is by no means clear that this phase of expansion was tied to the evolution of new adaptations to colder environments (Hoffecker 2005a:28–46).[4] Most human skeletal remains in Europe dating to 0.5–0.3 mya may be assigned to the taxon *Homo heidelbergensis*, representatives of which are also present in Africa at this time (Rightmire 1998). A 0.5-million-year-old tibia recovered from Boxgrove in southern England indicates limb dimensions that are comparable to those of modern tropical peoples and provides no evidence of anatomical adaptations to low temperatures (Stringer and Trinkaus 1999).[5]

Signs of a dietary shift or new technologies that might have facilitated expansion into higher latitudes also are scarce. A number of *H. heidelbergensis* sites, including Boxgrove (Parfitt and Roberts 1999), contain large mammal bones with stone-tool percussion and cut marks that reflect significant consumption of meat (Turner 1999), but not necessarily more so than at middle latitudes in Eurasia before 1 mya (Hoffecker 2005a:38–40).

Like their skeletal morphology, the stone tool technology of European hominids after 0.5 mya bears close resemblance to that of their contemporaries in Africa. Many sites contain hand axes and other Acheulean bifacial tools (Gamble 1986:141–158). Although isolated examples of nonstone implements have been found, such as the wooden spears or probes from Schöningen (Thieme 1977), such tools were also probably manufactured at lower latitudes. Controlled fire was long believed to represent an innovation for coping with northern climates[6] but has now been firmly identified in the Near East at ca. 0.79 mya (Goren-Inbar et al. 2004) and may have a much longer history of use in Africa (Brain and Sillent 1988).

Throughout the period between 0.5 and 0.3 mya, human settlement in northern Eurasia was almost entirely confined to western and central Europe. Some limited traces of occupation are present along the southern margin of eastern Europe (Hoffecker 2002a:42–48),[7] but no sites are found in northern Asia (i.e., above 45°N). The site at Longgushan in northern China (40°N) was occupied only during warm intervals between 0.67 and 0.41 mya (Boaz and Ciochon 2004:108–123). The pattern of settlement—in both space and time—apparently reflects the inability of *Homo heidelbergensis* and its East Asian contemporaries to cope with the low winter temperatures and the reduced biological productivity of northern Asia.

A possible exception to the pattern is the Siberian site of Diring, which is located 140 kilometers south of Yakutsk on the Lena River at latitude 61°N (Mochanov 1988). The open-air site is found on the highest terrace (about 120 meters above the modern river) and contains more than 4,000 chipped pebbles, flakes, and fragments of quartzite and other stone. The artifacts represent a lag deposit on a formerly deflated surface that is dated by luminescence to between 0.37 and 0.26 mya (Waters, Forman, and Pierson 1997).

Although some archaeologists have suggested that the artifacts at Diring were produced by geological processes (e.g., Goebel 1999:210–212), the investigators insist that they exhibit the characteristics of humanly flaked stone and comprise lithic reduction sequences (Waters, Forman, and Pierson 1997:1283; M. R. Waters, pers. comm.). Diring might reflect a temporary northward movement of people into central Siberia during the warm interglacial period dating to 0.34–0.30 mya (correlated with MIS 9), when climates were warmer than present. It deviates significantly, however, from the overall pattern of early human settlement before the late Pleistocene.[8]

Neandertals in Siberia

By 0.3–0.25 mya, the Neandertals (*Homo neanderthalensis*) had evolved in western Europe—almost certainly from the local *H. heidelbergensis* population—and they subsequently dispersed eastward into some of the colder and drier regions of northern Eurasia (Hublin 1998). The Neandertals became the first hominids to settle widely across eastern

Europe, and their sites are found not only in the southern margin but also on the central East European Plain as far as 52°N (Hoffecker 2002a:64–87). Most of these sites date to the Last Interglacial climatic optimum—between 128,000–116,000 years ago (MIS 5e age equivalent)—or later. The Neandertals apparently occupied sites in the Altai region of southwestern Siberia and may have visited a cave in the Upper Yenisei Valley in south-central Siberia (Goebel 1999:212–213).[9]

Unlike their predecessors, the Neandertals had evolved a suite of adaptive responses to cold environments, which were almost certainly tied to their colonization of eastern Europe and southwestern Siberia. Anatomical adaptations included short limbs relative to trunk length—and especially short distal limb segments—that would have reduced heat loss through the extremities. The Neandertals possessed thick chests and large heads, which further increased the ratio of their body mass to exposed surface area (Coon 1962:529–548; Trinkaus 1981; Holloway 1985:320–321) (figure 3.3).[10]

The Neandertal diet may have been even more important than anatomy as a cold-climate adaptation. Stable isotope analyses of their

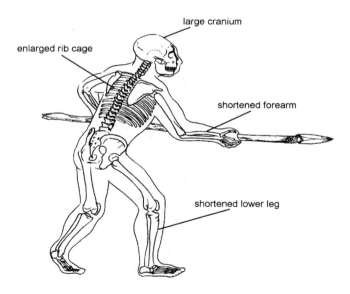

large cranium

enlarged rib cage

shortened forearm

shortened lower leg

Figure 3.3 Neandertal anatomical adaptations to cold climate (after Churchill 1998:47, fig. 1)

bones indicate protein derived almost entirely from mammals—even in western Europe during the Last Interglacial, when digestible plant foods were probably abundant (Bocherens et al. 1999). Many sites occupied by the Neandertals contain substantial quantities of large mammal remains (e.g., red deer, sheep, and bison), and where sufficiently well preserved, bone surfaces often exhibit traces of butchering (e.g., Baryshnikov, Hoffecker, and Burgess 1996). The characteristics of the assemblages— especially the recurring predominance of prime-age adults—suggest that most of the large mammals represent hunted rather than scavenged prey (Chase 1986; Hoffecker and Cleghorn 2000).[11]

Technology probably had a much lesser role in adaptation to cold environments among the Neandertals. Nevertheless, the development of hafting methods for spears and cutting/scraping implements increased the effectiveness and efficiency of tools and weapons (Shea 1988; Anderson-Gerfaud 1990). This would have improved their ability to inhabit less-productive environments with more-dispersed resources. The hafting techniques may have been tied to heavy use of Levallois prepared-core techniques, which allow superior control over the size and shape of stone tool blanks (Klein 1999:328).[12]

During the Last Interglacial climatic optimum, when temperatures in northern Eurasia were several degrees higher than those of today, the Neandertals expanded their range eastward into the upland Altai region of southwestern Siberia (51°N) (Goebel 1999:212). At the time, the area supported a coniferous forest with some broadleaf taxa and a variety of large mammals, including red deer, horse, and rhinoceros (Grichuk 1984; Derevianko et al. 1998:34–38). Human skeletal remains are confined to an isolated and nondiagnostic tooth from Denisova Cave (670 meters above sea level), but the associated Levallois Mousterian artifacts are similar to those produced by Neandertals in western Eurasia. Several isolated teeth from younger deposits in the Altai exhibit affinities to the Shanidar Neandertals (Turner 1990).

Neandertals also may have occupied the Upper Yenisei River in the south-central Siberia area during the Last Interglacial climatic optimum. Although no human skeletal remains have been recovered from Dvuglazka Cave, Levallois Mousterian artifacts were recovered in association with bones of large mammals that reflect warm conditions (Abramova 1981).[13] Taphonomic studies of the faunal remains are not available, but the economy—with its emphasis on the hunting of large mammals—may have been similar to that of the west Eurasian Neandertals.

The upper Mousterian levels in Denisova Cave, along with younger occupation layers at other sites in the region, indicate that the Neandertals continued to inhabit the Altai during the colder periods that followed the Last Interglacial (Abramova 1989:156–167; Derevianko et al. 1998; Goebel 1999:212–213). Neandertals may have been present not only during the interstadial that correlates with MIS 3 but also during the preceding cold phase (MIS 4) or Lower Pleniglacial, when temperatures were significantly lower than the present day.[14] The later occupations of the Altai—where January mean temperatures may have been as low as −20°C—would seem to represent the maximum cold tolerance of the Neandertals (Hoffecker 2005a:57–59).

Origins and Dispersal of Modern Humans

Modern humans (*Homo sapiens*) evolved in Africa and subsequently dispersed across Eurasia and into other parts of the world. They repeated the same "out of Africa" pattern followed by early humans (Stringer and Gamble 1993:214–216; Klein 1999:503–505), but in a much shorter period of time and without differentiation into several species (e.g., *Homo erectus, H. neanderthalensis*). Modern humans also extended the range of *Homo* beyond that of their predecessors into colder environments where winter temperatures fell far below −20°C (Hoffecker 2005a:81–82). They seem to have been the first hominins (and hominids) to occupy habitat above 60°N and were thus the only humans with the potential to inhabit Beringia (Hoffecker and Elias 2003).

Like the Neandertals in Europe, modern humans evolved roughly 300,000–200,000 years ago from local representatives of *Homo heidelbergensis* or a closely related form (Rightmire 1998). New dates on a modern skull from Omo-Kibish (Ethiopia) indicate that *Homo sapiens* was present in Africa by at least 195,000 years ago (McDougall, Brown, and Fleagle 2005). The most significant anatomical change seems to have been an expansion of brain volume (to 1,350 cubic centimeters or more) and development of more vertically shaped frontal bone (Klein 1999:498–502). Equally, if not more, important is archaeological evidence of major changes that are often characterized as "behavioral modernity" (e.g., d'Errico 2003). These behavioral changes are inferred from evidence for long-distance transport of materials, application of mineral pigment, use of grinding stones, regional variation in artifact styles, production of bone implements, personal ornamentation, and various forms of art and ritual (McBrearty and Brooks 2000).

Some paleoanthropologists believe that spoken language was the centerpiece of behavioral modernity (e.g., Mellars 1996:387–391; Klein 1999:515–517). Documenting the presence or absence of spoken language (specifically syntactical language)[15] has been problematic, however (d'Errico et al. 2003), although there is some evidence for a genetic mutation related to speech after 200,000 years ago (Enard et al. 2002). It is apparent that by 100,000 years ago, African *Homo sapiens* had evolved the capacity for communicating or projecting mental constructs through a variety of media, which, by implication, probably included spoken syntactical language (Davidson and Noble 1989; Bickerton 1990). Equally important may be recursion—the ability to generate a potentially infinite array of combinations from a finite set of elements—which has been described as the "core property" of language (Hauser, Chomsky, and Fitch 2002:1571). Recursion is manifest in the archaeological record—especially with respect to art and technology—created by modern humans after 100,000 years ago.

Behavioral modernity seems to have been the critical factor that permitted modern humans to rapidly colonize a wide variety of habitats outside Africa at some point after 0.1 mya (Fagan 1990; Gamble 1994:181–202). Modern humans adapted to these habitats—which included cold steppe and desert—with novel and often complex technologies that are documented in the archaeological record for the first time. At higher latitudes, technology for cold protection in the form of tailored clothing and artificial shelters was especially important (Hoffecker 2002a:158–161). Significantly, modern humans in these areas before 20 cal ka retained an anatomy more suitable for tropical climates and presumably inherited from their African ancestors (Trinkaus 1981; Holliday 1999). In environments where resources were widely dispersed, new technologies that either improved the efficiency of large-mammal hunting or broadened the diet range to small mammals, birds, and other less-accessible prey also were important (Klein 1999:535–544; Hoffecker 2005b) (figure 3.4).

In addition to novel technologies, modern humans also probably created new organizational structures that helped them cope with very cold

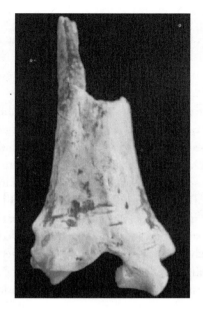

Figure 3.4 Tibia of hare exhibiting cut mark recovered from Kostenki 14 and dated by about 40 cal ka (photograph by JFH); the large quantity of small mammal remains in some early Upper Paleolithic occupations in northern Eurasia suggest an expansion of dietary breadth, probably through the design of new technologies for procuring prey animals that were inaccessible to the Neandertals

and arid environments (Gamble 1994:167–170). One of the more striking changes that modern humans brought to the archaeological record of Eurasia is the quantum increase in the maximum distances of raw materials such as stone and shell transported from their sources—up to and exceeding 700 kilometers (Gamble 1999:319–321). The pattern suggests the development of social networks spread across large areas, which—as among recent foragers—could have offset the problem of scarce and unpredictable resources (Stringer and Gamble 1993:212–213; Kelly 1995). And by 30–25 cal ka, there is evidence of at least temporary aggregations of people who may have collectively harvested concentrations of prey (e.g., Svoboda, Lozek, and Vlcek 1996:146–153; Hoffecker 2002b:125–129).

Although there are remains of *Homo sapiens* in the Levant as early as 100–90 cal ka,[16] colonization of most of Eurasia (and Australia) does not seem to have taken place until after ca. 60–50 cal ka (Stringer and McKie 1996:160–163). There is some evidence for an early phase of colonization in the tropical zone of Eurasia before 50 cal ka (Lahr and Foley 1994; Shen et al. 2002), and modern humans reached Australia by at least 50 cal ka (Bowler et al. 2003). The initial appearance of modern humans in northern Eurasia probably occurred before 40 cal ka. Significantly, the oldest sites appear to lie on the East European Plain and southern Siberia—the coldest and least productive areas above latitude 45°N. The earliest occupations in eastern Europe date to at least 40 cal ka and possibly as much as 45 cal ka (Sinitsyn 2002; Anikovich 2003),[17] and the oldest sites in Siberia probably date to 40 cal ka or more (Abramova 1989; Derevianko et al. 1998:106–128; Goebel 1999:213–214).[18]

Modern human settlement of northern Eurasia took place during the later phases of the interstadial that correlates with MIS 3. Several brief warm oscillations (Dansgaard-Oeschger Events) are recorded in Greenland ice cores between 50 and 30 cal ka (Dansgaard et al. 1993), and climates apparently were quite mild during these events. At times, temperatures in Siberia may have been close to those of the present day (and this period is sometimes locally characterized as an "interglacial" rather than an intersta-dial [e.g., Tseitlin 1979]). Open forest of pine and birch—interspersed with some steppic flora—covered much of southern Siberia and supported a variety of large mammals (Giterman and Golubeva 1967; Kind 1974). The timing and distribution of modern human settlement above latitude 45°N probably was influenced by oscillating climates during 50–30 cal ka; this is almost certainly the case with respect to occupation of areas beyond 60°N, which included at least a portion of western Beringia.

Modern Humans in Northern Asia

Broadly speaking, modern humans occupied Beringia as part of their wider colonization of northern Asia about 50–40 cal ka. Although permanent settlement of Beringia did not occur until changes in climate and vegetation took place at the beginning of the Lateglacial (16–15 cal ka), it was only possible in the context of modern human abilities to restructure the environment through novel and complex forms of technology and organization. Permanent occupation of Beringia was delayed by one or more factors, which may have included the primitive state of technology before 15 cal ka—modern humans gradually accu-mulated technological knowledge after 50 cal ka—and possibly the retention of their tropical zone anatomy (Hoffecker and Elias 2003:38).

Modern humans probably arrived in Northeast Asia about 45–40 cal ka but, as in Europe, the oldest occupations lie at or beyond the effec-tive range of the radiocarbon method and are difficult to date. More-over, modern human skeletal remains have not yet been recovered from the earliest sites, and their presence is inferred from the character of the archaeological finds in these sites (Goebel 1999:215).[19] Occupation layers containing stone blades are dated to 45–40 cal ka at Kara-Bom (Altai region) and Makarovo 4 (Upper Lena River) but otherwise lack archaeological evidence of modern behavior (Derevianko et al. 1998:106–128; Brantingham et al. 2001:736–737).[20] However, somewhat younger occupations—dated to 35–30 cal ka—contain ornaments, bone tools,

Figure 3.5 Siberia, showing the locations mentioned in the text

and possibly art, such as Layers 11–9 at Denisova Cave (Altai) and the open-air site of Tolbaga (Transbaikal region) (figure 3.5).[21]

The sites occupied in Northeast Asia during 45–30 cal ka are found below latitude 55°N in a variety of topographic settings, including caves and small open-air settings. Some sites are relatively large with debris concentrations, hearths, pits, and occasional traces of former dwellings (Goebel 1999:213). Evidence for cold-protection technology in the form of artificial shelters—huts or tents with interior hearths—and needles for sewn clothing (figure 3.6) are reported from Tolbaga (Vasil'ev, Kuznetsov, and Meshcherin 1987; Goebel 2001a:184). Eyed needles are also reported from Layer 11 at Denisova Cave but might be intrusive from a younger level (Derevianko et al. 1998:41–42). The role of technology in the colonization of northern Eurosia is discussed in box 3.1

Faunal remains from these sites represent a variety of large mammals, including horse, steppe bison, reindeer, goat, sheep, woolly rhinoceros, and others (Goebel 1999:214–215). Unlike eastern Europe, however, there is limited evidence in Siberia before 30 cal ka for harvesting small mammals, birds, or fish.[22] Perhaps the occupants of these sites lacked the technology—such as darts, nets, snares, and so forth—required for obtaining such prey. It should be noted that Siberian sites before 30 cal ka also lack evidence for long-distance transport of materials (Goebel 1999:214), suggesting that exchange and alliance networks were limited.

At times during the later phases of the MIS 3 interstadial, modern human groups extended their range into latitudes above 60°N. The site of Mamontovaya Kurya, located at 66° 34′ N in Russia, yielded a small assemblage of stone artifacts and an incised mammoth tusk dating to roughly 40 cal ka (Pavlov, Svendsen, and Indrelid 2001). And, as noted

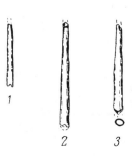

Figure 3.6 The oldest known eyed needles, recovered from Kostenki 15 on the Don River in Russia and dated to roughly 35–30 cal ka (after Rogachev and Sinitsyn 1982:170, fig. 59)

Box 3.1

Prelude to Beringia: Modern Human Technology and the Colonization of Northern Eurasia

Archaeological evidence indicates a capacity for technological innovation and design complexity among anatomically modern humans (Klein 1999:535–544) that probably was essential to their rapid spread into a wide range of habitats and climate zones, including the very cold environments of Northeast Asia, and therefore was critical to the settlement of Beringia (Turner 2002:142–146; Hoffecker 2005a:89–91). Identifying when and where specific technological innovations were developed—and reconstructing the details of their design—in the archaeological record of 45–15 cal ka is a challenge.

The invention of new technology for cold protection—sewn tailored clothing and artificial shelters—was especially important to the colonization of higher latitudes (Fagan 1990:158–160). Neither seems to have been developed by the Neandertals, which may have been the primary constraint on their ability to inhabit Northeast Asia. Evidence for sewn clothing in the form of eyed needles of bone and ivory has been found in sites occupied by modern humans in eastern Europe (Kostenki) and southern Siberia (Tolbaga) as early as 35–30 cal ka (figure 3.6). Traces of artificial shelters with interior hearths have been recorded at several sites in Europe and Siberia dating to 30 cal ka (e.g., Barca II [Slovakia]) (Troeng 1993; Hoffecker 2005b). The lack of evidence for such technology in the oldest modern human occupations may simply be a function of poor preservation and limited sampling.

Also critical to the occupation of cold and arid environments was novel food-getting technology that permitted modern humans to expand their niche to previously inaccessible resources and to increase their foraging efficiency. This novel technology is inferred from the presence of large quantities of small mammals (such as hare) in some north Eurasian sites as early as 40 cal ka (Stiner et al. 1999; Hoffecker 2005a:90), and also from evidence of heavy freshwater aquatic food consumption derived from stable isotope analyses of human bones found in sites occupied at 30–25 cal ka (Richards et al. 2001). Younger sites in Europe contain more direct evidence: traces of nets, remains of throwing darts, and possible trap components (Pidoplichko 1976; Soffer 2000). Innovative mechanical weaponry, such as the spear-thrower (19 cal ka) and bow and arrow (14 cal ka), improved foraging efficiency (Cattelain 1997).

Variation in mtDNA sequences among modern dogs suggests that domestication (biotechnology) took place in Asia after the Last Glacial

Maximum (LGM) (Savolainen et al. 2002), and skeletal remains of domesticated dogs are reported from Bonn-Oberkassel (Germany) and Eliseevichi (Russia) at 18–16 cal ka (Benecke 1987; Sablin and Klopachev 2002). Christy G. Turner II (2002:145) argues that domesticated dogs were critical for the successful occupation of Beringia (as was tailored clothing). Another technological innovation of modern humans that might have been essential to colonization of Beringia is watercraft (Dixon 2001:294–295), which was almost certainly used to colonize Australia and parts of Oceania at 50–30 cal ka.

in chapter 1, a site in Northeast Asia (Yana RHS)—near the mouth of the Yana River at 71°N—recently was dated to almost 30 cal ka (Pitul'ko et al. 2004). Because it lies east of the Verkhoyansk Mountains, Yana RHS represents the earliest known occupation of Beringia (as defined in chapter 1).

It remains unclear whether these sites were occupied by people who were present at high latitudes throughout the year or only during the warmer months. It is apparent, however, that they reflect the relatively warm climates of the later MIS 3 interstadial and their effects on biota. Traces of high-latitude settlement are absent during the subsequent cold period (MIS 2 age equivalent). Palynological evidence suggests that larch forests nearly reached their modern limits in western Beringia during 42–36 cal ka (Anderson and Lozhkin 2001). Cooler and drier conditions followed from about 36 to 33 cal ka, marked in western Beringian pollen records by the appearance of herb tundra or birch-shrub tundra in regions previously occupied by larch forest (Brigham-Grette et al. 2004). At the end of the interstadial—ca. 35–30.5 cal ka—there was a brief warm interval marked by the return of larch and birch forest tundra to the Yana-Indigirka-Kolyma lowlands.

Yana RHS may have been occupied during the final MIS 3 warm interval. The artifacts comprise unifacial and bifacial stone tools, along with several bone, ivory, and horn items, that were deposited in stream sediments dated to about 27,000 radiocarbon years BP (slightly less than 30 cal ka). Associated mammal remains include reindeer (most common), mammoth, horse, bison, and hare (*Lepus tanaiticus*); additional remains probably associated with the occupation include musk ox, wolf, polar fox, and some unidentified birds (Pitul'ko et al. 2004) (Figure 3.7).

Pitul'ko et al. (2004:55) believe that the bird remains represent an important part of human hunting activity at Yana RHS, which suggests that the site was occupied during the warm season. In historic times, groups of Yukaghir moved north into the same area to fish and snare migratory birds in the spring (Forde 1934:103–105). Both the bird and small mammal remains at Yana RHS may indicate that, contrary to the pattern seen in the southern sites, modern humans in Siberia were expanding their dietary range with various novel technologies at this time. It also suggests that the Siberians may have been moving over distances of hundreds of kilometers to exploit seasonal concentrations of prey animals during the year. Both are characteristic of recent hunter-gatherers in northern interior environments (Kelly 1995).

At present—and in the absence of any evidence for other sites in Beringia until after 16 cal ka—we conclude that Yana RHS represents part of the temporary, and probably seasonal, movement above 60°N that took place across Eurasia during one or more of the warm intervals of the later MIS 3 interstadial.[23] In addition to significantly milder

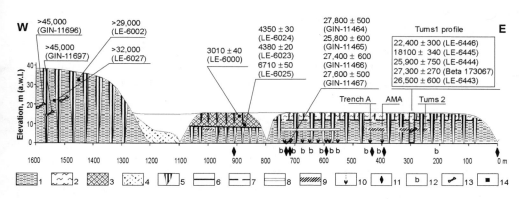

Figure 3.7 Yana RHS ("Rhinoceros Horn Site"): earliest dated site within the boundaries of Beringia, as defined in this book, located on the lower Yana River at latitude 71°N (reprinted with permission from Pitul'ko et al. 2004:53, fig. 3, copyright 2004 AAAS.)

climates, wood fuel in the form of shrubs and trees—perhaps the most important variable—was abundant in western Beringia during these intervals. In contrast, there is no compelling evidence thus far for occupations in eastern Beringia (or elsewhere in the New World)[24] at this time. The pattern might reflect the less-favorable conditions in eastern Beringia (see chapter 2) or the lack of a connecting land bridge during the warmer intervals of MIS 3, or both. Permanent year-round settlement of Beringia (and latitudes above 60°N) apparently awaited the beginning of the Lateglacial at 17–16 cal ka.

The Last Glacial Maximum

Temperatures declined throughout the Northern Hemisphere after 28–27 cal ka, inaugurating the early phases of the Last Glacial Maximum (LGM). After a brief mild oscillation at about 25 cal ka—which correlates with a weakly developed soil in many parts of northern Eurasia—temperatures declined further, reaching a cold maximum at 23–21 cal ka. Blankets of loess were deposited in many parts of northern Eurasia at this time. During the peak cold phase, the January mean in southern Siberia is estimated at 6–10°C lower than today, and precipitation fell by roughly 10 percent (Velichko 1984:275–283).[25]

Despite the increased cold and aridity—or perhaps because of it—modern human populations in Siberia thrived during the early phases of the LGM in southern Siberia. Owing to the aridity of the interior, most lowland areas remained unglaciated. Climate conditions and loess deposition favored reduced tree cover and expanded steppic vegetation, supporting increased numbers of mammoth, horse, bison, and other large-mammal grazers. Faunal remains found in archaeological sites occupied during this interval also contain numerous small mammals and birds (Goebel 1999:216–218).

Upper Paleolithic sites in southern Siberia dating to 27–23 cal ka contain traces of semi-subterranean dwellings, some storage pits, and substantial quantities of occupation debris. They include the famous sites of Mal'ta and Buret' located west of Lake Baikal, as well as other sites in the Yenisei Valley, Upper Lena Basin, and Transbaikal region (Abramova 1989; Abramova et al. 1991; Goebel 2001b). The occupants of these sites apparently pursued a broad-based northern interior economy, harvesting a variety of mammals and birds (Ermolova 1978:15–29). Analysis of stable isotope values for a human bone fragment from Mal'ta indicates that 25–50 percent of the diet was derived from freshwater aquatic sources (Richards et al. 2001:6529), and many of the carved figurines from this site represent waterfowl (Gerasimov 1964). The large dwellings and associated pits at Mal'ta and other sites may reflect mass harvesting of temporary prey concentrations and long-term food storage (Goebel 1999:216) (Figure 3.8).

Several smaller sites may represent short-term camps used for a more limited range of activities. An example is Tomsk—located on an Ob River tributary in the West Siberian Lowland—which appears to be a mammoth kill-butchery site (Petrin 1986:72–74; Abramova 1989:174–175). However, none of the early LGM sites has been found above 60°N, and it is unclear whether the Siberians were continuing to occupy high-latitude areas such as the Lower Yana River—at least on a temporary seasonal basis—as during the preceding period. The lack of known sites at higher latitudes may simply reflect their low archaeological visibility.

The number of occupations dating to the maximum cold peak at 23–21 cal ka is small, and much of Siberia may have been abandoned at this time (Goebel 1999:218). A similar pattern is apparent on the East European Plain (Soffer 1985; Dolukhanov, Solokoff, and Shukurov 2001:709, fig. 6; Hoffecker 2002a:200–201). Recently, Y. G. Kuzmin and S. V. Keates (2005:785, table 3) noted that a number of occupations

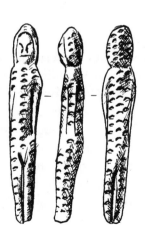

Figure 3.8 Carved figurine from Buret in southern Siberia dated to about 24 cal ka confirms production of tailored fur clothing, including hoods (after Chard 1974:27, fig. 1.13)

yield radiocarbon dates in this range, and they argued that evidence for abandonment of Siberia during the maximum cold interval was lacking. Firmly dated sites in the 23–21 cal ka range above latitude 50°N remain scarce, however,[26] and indicate a probable decline in settlement for at least a thousand years (Goebel 2002:121, fig. 9.3).

It is not entirely clear why modern humans might have abandoned parts of northern Eurasia during the cold maximum. Food sources in the form of large mammals were present in most periglacial areas—supported by steppic plants that apparently thrived on nutrient-rich loess—and this is especially well documented in Beringia (see chapter 2) (Guthrie 1990; Walker et al. 2001). Although mean winter temperatures probably fell below –30°C in most of these areas, recent hunter-gatherers equipped with similar technology, such as the Yukaghir in the Upper Yana River, endured even lower January means (e.g., Stepanova, Gurvich, and Khramova 1964).

There are several possible explanations for the pattern. First of all, the technology for cold protection developed in northern Eurasia about 23 cal ka—even if broadly comparable to that of recent foraging peoples—may not have been as effective (and the differences not apparent in the archaeological record). Second, although fresh bone was almost certainly available as a fuel substitute for wood—and was used heavily on the central East European Plain at other times (Hoffecker 2002a:200–212)—adequate starter fuel in the form of woody shrubs[27] may have been scarce in some places. Finally, the retention of body dimensions similar to those of peoples in the tropical zone (Holliday 1999) probably exposed the northern Eurasians to high rates of cold injury, despite their fur clothing (Hoffecker and Elias 2003:41; Hoffecker 2005a:94–95).[28]

In any case, while the absence of known sites in Beringia during the LGM might reflect reduced archaeological visibility in remote places like Chukotka and Alaska, the pattern is consistent with the overall decline in settlement in northern Eurasia above latitude 50°N at this time. Although Colinvaux and West (1984) suggested that lack of adequate food resources could explain the absence of settlement in Beringia during the LGM, this does not appear to have been the critical factor because large mammals continued to occupy unglaciated high latitude areas, including Beringia (Matthews 1982:139–142). The explanation probably lies among the above-discussed reasons for the general scarcity of occupation in these areas during the cold maximum.

Reoccupation of Northeast Asia

The number of dated occupations in southern Siberia begins to increase slowly after 21–20 cal ka (ca. 18,000–17,000 radiocarbon years BP), and the pattern more or less follows the rising temperature curves—based on various proxy data sources—in the millennia after the cold maximum. Most occupations date to between 19 and 13 cal ka (Goebel 2002:121) and include open-air sites like Chernoozer'e (West Siberian Lowland), Kokorevo and Afontova Gora (Upper Yenisei Valley), Verkholenskaya Gora (Upper Angara Valley), and Studenoe (Lake Baikal region) (Abramova et al. 1991; Goebel et al. 2000; Vasil'ev 1996, 2001). All of them are found below latitude 60°N.

The pattern of settlement in southern Siberia after 20 cal ka was strikingly different from that of the period before the cold maximum (Goebel 2002; Hoffecker 2005a:110–112). The sites contain relatively small and thin occupation layers, but traces of semi-subterranean dwellings are absent and pits are rare. The faunal assemblages are often dominated by one mammal taxon (e.g., reindeer or red deer). Small wedge-shaped stone cores are common and were used for mass production of microblades that were often inset into the slotted margins of bone points. Examples of slotted bone points containing one or more rows of inset microblades have been found at Kokorevo I and Chernoozer'e II (Abramova 1979:110, fig. 53; Gening and Petrin 1985:48, fig. 17). Microblades were later mass-produced in Beringia and probably used in a similar way.

The sites apparently reflect a highly mobile settlement system, which has been characterized by Ted Goebel as a "microblade adaptation" (Goebel 2002:123–126). They seem to have been occupied briefly by small groups for a limited range of activities—often focused on the hunting of reindeer or red deer (*Cervus elaphus*). The technology was based on a very economical use of raw material, and the portable microcores were sometimes reused as tools. The conditions that had permitted harvesting and storing of large quantities of food—and thus extended settlements—during the early LGM were apparently no longer extant. This may have been a consequence of expanding boreal forest, which would have reduced the density of food resources.

Sites dating to 19–13 cal ka contain, in addition to large mammals such as reindeer and red deer, some remains of small mammals and other vertebrates that suggest a continuing trend toward a broad-based northern interior economy (Goebel 1999:219–220). Hare (*Lepus* sp.) is especially common at some sites such as Tashtyk I in the Upper Yenisei Valley (Abramova et al. 1991:30, table 1). Bird remains are present at many localities, and fish bones appear in sites occupied after 15 cal ka (Hoffecker 2005a:111–112). Traces of mammoth are largely confined to tusk fragments and may represent scavenging of remains for ivory (Goebel 2002:126). All of these trends seem to anticipate the postglacial economy that developed with the settlement of Beringia.

The same variables that may have had a role in the retreat from higher latitudes during the LGM are pertinent to the reoccupation of Northeast Asia after 20 cal ka. Some major innovations are evident at this time in other parts of Eurasia, and technological improvements in Siberia—such as better insulated clothing[29] and others not especially visible in the archaeological record—could have significantly enhanced the ability of its inhabitants to survive in high-latitude environments (Hoffecker and Elias 2003:38). The spreading availability of wood fuel might have been critical. In contrast to the East European Plain, evidence of bone fuel is not common in the Siberian sites (although an occupation level at Krasnyi Yar I in the Upper Angara River that dates close to the cold maximum yielded traces of combustible shale in former hearths [Goebel 1999:218]). As the Siberians resettled northern areas after 20 cal ka, they seem to have been dependent on wood fuel and probably followed the gradually expanding tree line northward.

A parallel reoccupation of areas abandoned in Europe took place after 20 cal ka, and the European sample of human skeletal remains dated to 20–12 cal ka reveals shorter limbs (Holliday 1999). This would have reduced susceptibility to cold injury and might have been a factor in reoccupation of the central plain. As in the case of earlier periods, however, the Siberian sample is too small and fragmentary to determine whether similar anatomical changes occurred in northern Asia (Alekseev 1998; Hoffecker and Elias 2003:38).

By 15 cal ka, people were present in the middle Lena Basin and were living—apparently on a year-round basis—at the doorstep of Beringia. Dyuktai Cave, which is located on the Aldan River at latitude 59° 18' N, was occupied at 15 cal ka or perhaps slightly earlier, by makers of wedge-shaped microblade cores, burins, bifacial knives, and other stone implements (Mochanov 1977:26–31; Mochanov and Fedoseeva 1996a:167–174). Although the artifacts are similar to those of the more southern Siberian sites, Mochanov (1969) defined a new archaeological culture (Dyuktai culture) on the basis of these assemblages (figure 3.9).

The faunal remains recovered from the lower layers of Dyuktai Cave indicate an economy that was developing in the same direction as the south Siberian sites. Even in the oldest occupation level, traces of mammoth are almost entirely confined to tusk fragments (presumably scavenged). Although some horse and bison remains are present, hare is the most common taxon in the lowest two levels (Layers IX–VIII), and both birds and fish are represented in these layers. In a younger level (Layer VIIa), reindeer and moose (*Alces alces*) become more common. The contents of the hearths reveal that wood was available for fuel (Mochanov and Fedoseeva 1996a:167–174).

It is the Dyuktai culture that appears to have provided the source for the earliest permanent settlement of Beringia, which took place shortly after the cave was initially occupied. As described in chapter 4,

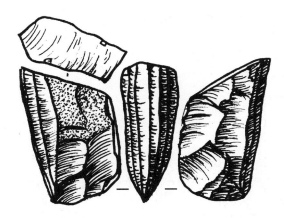

Figure 3.9 Wedge-shaped microblade core from Layer VIIa at Dyuktai Cave, which is dated to roughly 15 cal ka (after Mochanov 1983:271, fig. 169)

the oldest firmly dated in sites in Beringia (roughly 15–14 cal ka) contain similar microblade cores, microblades, burins, and evidence of bifacial tools (Holmes 2001; C. E. Holmes, pers. comm. 2006). They also reflect a similar—although rapidly evolving postglacial—northern interior economy (Yesner 2001). This seems to link the Lateglacial settlement of Beringia to the broader reoccupation of interior Northeast Asia after the LGM.

Nevertheless, the settlement of Beringia coincides with major changes in climate and biota that apparently removed remaining barriers to occupation at the beginning of the Lateglacial. The principal barrier may have been related to the availability of fuel in the form of woody shrubs—especially willow (Guthrie 1990:277). The latter spread across Beringia 16–15 cal ka as mesic shrub tundra replaced more steppic environments in the lowlands and somewhat later at higher elevations (e.g., Bigelow and Powers 2001). The distribution of sites occupied by humans after 15 cal ka seems to follow the same pattern.

The Beginning of the Lateglacial

Based on available evidence, permanent settlement of Beringia, including eastern Beringia, took place at the beginning of the Lateglacial intersta-dial. Archaeological sites dating between 15 and 13.5 cal ka (15,000 and 13,500 calibrated years ago) are found in various parts of Beringia, al-though the status of many remains problematic. The earliest firmly dated sites of this interval are found in the Tanana Basin of central Alaska. The site of Berelekh—located near the mouth of the Indigirka River in the northwest lowlands of Beringia—probably dates to the earlier phase of the Lateglacial as well.

The appearance of these sites coincides with major changes in climate and biota that occurred at the base of the Birch Zone (Hopkins 1982:10–11). Both summer and winter temperatures rose significantly after 15 cal ka (Sher et al. 2002) (figure 4.1), as mesic shrub tundra veg-etation spread across the lowland areas of Beringia. Birch—in the form of dwarf birch (*Betula nana*)—increased dramatically in numbers, but willow (*Salix* sp.) also became more common at this time. As steppe-tundra plant communities shrank, many of the grazers that had domi-nated the large mammal fauna of the LGM declined, departed Beringia, or became extinct (Matthews 1982:140; Guthrie 2003, 2004, 2006:207, fig. 1).

Despite the rise in moisture, which was a critical variable in the transition to shrub tundra (Guthrie 2001), the Bering Land Bridge remained largely intact during the earlier Lateglacial (Elias et al. 1996). Perhaps reduced ice cover in the Bering Sea provided the increased available moisture (see chapter 2). In any case, humans reentered Beringia during a twilight period before rising sea levels caught up with the rapid pace of climate change and its effects on plants and animals. The Lateglacial Beringians found themselves living in a postglacial world of shrub tundra.

The sites in Beringia that antedate 13.5 cal ka are small and yield modest quantities of artifacts and debris; they lack traces of dwellings or pits. Nevertheless, artifacts recovered from a well-dated context in central Alaska during 2003–2005 reveal a close affinity with the industry from Dyuktai Cave described in chapter 3. Wedge-shaped microblade cores, microblades, and burins in the lowest cultural level at Swan Point indicate

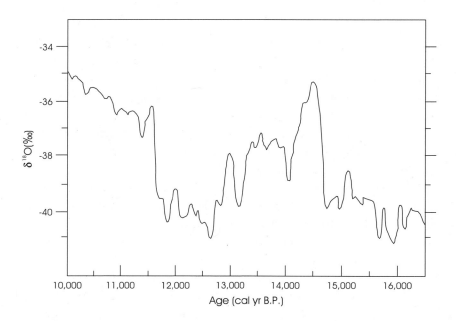

Figure 4.1 Climate change in the Northern Hemisphere between 16 and 10 cal ka: GISP 2 ice core record

that Lateglacial Beringia probably was occupied by people from the adjoining Lena Basin (Holmes and Crass 2003; C. E. Holmes, pers. comm., 2006). And even the earliest sites contain evidence for long-distance movement of raw materials—specifically obsidian (Holmes 2001:167).

From the outset, the Beringian economy exhibits many similarities to that of later northern interior peoples (Cook 1969:278). The large mammal prey were those of the postglacial epoch, such as elk, moose, sheep, and caribou. Traces of mammoth are primarily confined to tusk fragments that may have been scavenged from natural accumulations (Holmes 1996:317), although mammoth were not yet extinct in Beringia (Guthrie 2004, 2006). Small mammals—especially hare—were harvested in significant quantities, and there was a particularly heavy emphasis on birds—especially waterfowl (Yesner 2001:321–322). The latter presumably reflects considerable expansion of waterfowl summer habitat at this time.

What triggered the movement of people into Beringia at the beginning of the Lateglacial period? The close association of the sites with the spread of shrub tundra suggests a connection, and some have argued that the significant increase in woody plants—a fuel source— was the critical factor (e.g., Guthrie 1990:277; Hoffecker, Powers, and Goebel 1993:48). Most of the increase during the Lateglacial was in the form of woody shrubs and not full-sized trees (see chapter 2).

While in some places, shrubs such as willow (and occasional poplar trees) might have offered sufficient fuel, Beringian archaeological sites consistently yield evidence for the burning of bone (e.g., Cook 1969:242–243; Mochanov 1977:79; Goebel, Waters, and Dikova 2003:503; C. E. Holmes, pers. comm. 2005). Supplies of flammable bone were abundant, particularly in lowland basins where large—often frozen fresh— quantities of bone accumulated at stream confluences during the LGM (Péwé 1975:98), but they would have been hard to fire up owing to their high ignition temperature (Théry-Parisot 2001). The widespread availability of woody shrubs in lowland areas after 15–14 cal ka may have provided the needed ignition source (Hoffecker and Elias 2005).

Western Beringia

The mildest climates in western Beringia are found today in the southwest region—the southern coast along the Sea of Okhotsk and the

Kamchatka Peninsula. Conditions here are moderated by the Pacific Ocean, and especially by the warm currents that flow clockwise along its northwest rim. The southwest region is therefore the area that might be expected to yield traces of the earliest settlement of Beringia. It is also the region that was closest to existing human settlement in both interior Siberia—the Lena Basin—and the northern Far East.

For many years, in fact, the oldest occupation of Beringia was thought to be located in central Kamchatka at Ushki (latitude 56°N) in one of the most biologically productive areas of Beringia. In the early 1960s, the late N. N. Dikov discovered Paleolithic remains on Great Ushki Lake, and the lowest level at Ushki I was eventually dated to ca. 16.8 cal ka (14,000 radiocarbon years BP) (Dikov 1979). Moreover, this level contained an assemblage unlike any known at the time in Beringia—lacking microblades—that seemed consistent with its unique position in the chronology. But renewed research at Ushki in 2000 revealed that the oldest horizon probably was not occupied until about 13 cal ka (Goebel, Waters, and Dikova 2003) during a later phase of Beringian prehistory. The Ushki sites remain important, nonetheless, and are discussed in chapters 5 and 6.

Several other sites in the southwest region previously have been viewed as possible traces of early settlement in Beringia (Goebel and Slobodin 1999; Slobodin 2001), but these also appear to have been occupied at later times. The most notable of these is Kukhtui III, which is located on a small river near the coast of the Sea of Okhotsk. Mochanov (1977:87–90), who discovered this site in 1970, suggested that the lowermost component—associated with pollen-spore samples that indicate cold climates—dated to the terminal Pleistocene (Mochanov and Fedoseeva 1996b:224). As in the lowest level at Ushki, microblade technology is absent. However, the lowermost component at Kukhtui III remains undated, and most archaeologists believe that it is of Holocene age (e.g., Dikov 1979:103–104; Goebel and Slobodin 1999:107).[1]

BERELEKH AND THE MAMMOTH "CEMETERY"

As a result of the redating of the earliest horizon at Ushki, the site of Berelekh remains the oldest known post-LGM occupation in western Beringia. Although some uncertainty continues to surround its dating, it is the only such site between the Lena Basin and the Bering Strait that apparently antedates 13.5 cal ka. Confounding the expectation that people would move first into the milder southwest region, Berelekh is found in the northwest lowlands of Beringia—near the Arctic Ocean coast above latitude 70°N. Nevertheless, a dated core from Dolgoe Lake (roughly 600 kilometers west of the site) confirms a sharp rise in birch pollen at 14 cal ka (Pisaric et al. 2001:238–239), and the occupation layer is filled with fragments of woody shrubs (Figure 4.2).

As in the case of many Paleolithic open-air sites on the East European Plain (Klein 1973:52–54; Hoffecker 2002a:34), Berelekh was identified through the discovery of mammoth bones and tusks. The latter had been known to local inhabitants for years but were not reported in a publication until 1957. In 1970–1971 and 1980, N. K. Vereshchagin, G. F. Baryshnikov, and others investigated the locality, which is found on the upper reaches of a small tributary of the Indigirka River. With

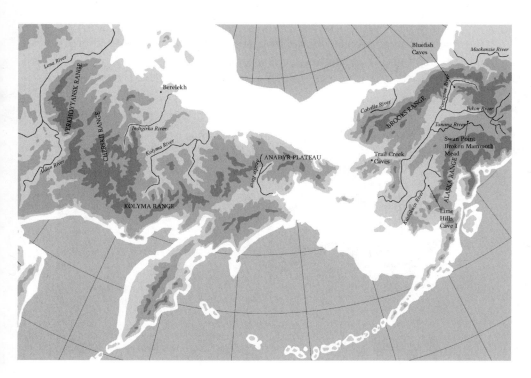

Figure 4.2 Beringia during the beginning of the Lateglacial (15–13.5 cal ka), showing the location of sites mentioned in the text

the use of pressure hoses on the frozen sediment, they recovered more than 8,800 bones, teeth, and tusks of mammoth (*Mammuthus primigenius*) representing a minimum of 156 individuals, along with a small number of remains of other vertebrates (including steppe bison, reindeer, and ptarmigan). Also recovered were mammoth soft tissues, including 2 kilograms of hair and a hind leg (Vereshchagin 1977:17–31; Vereshchagin and Ukraintseva 1985:105) (Figure 4.3).

Radiocarbon dates on the mammoth remains yielded calibrated estimates of 16 and 14 cal ka, and the associated wood fragments produced seven ages between 16 and 12 cal ka (Mochanov 1977:78–79). While the range of dates is broad, most are older than 13.5 cal ka (table 4.1).

Although the concentration of mammoth bones at Berelekh is described as a *kladbishche* or "cemetery" (e.g., Mochanov 1977:76), the term has always been placed in quotation marks and the process responsible for their accumulation is a matter of debate (Goebel and Slobodin 1999:120). The bones were recovered from bedded sandy loam apparently deposited as floodplain alluvium or possibly lake sediment. They might have been derived from natural mortality and concentrated by

Figure 4.3 Berelekh: washing mammal bones out of frozen silt in 1971 (photograph courtesy of G. F. Baryshnikov)

Table 4.1

Calibrated Radiocarbon Dates from Sites in Beringia: Earlier Lateglacial Interstadial (ca. 15–13.5 cal ka)

Site	Level[a]	Material Dated	Lab No.	Uncalibrated Date	Calibrated Age[b]
Berelekh	Cultural layer	Wood	LE-998	10,600 ± 90	12,518 ± 168
	Cultural layer	Wood	GIN-1021	12,930 ± 80	15,810 ± 502
	Cultural layer	Wood	IM-152	13,420 ± 200	16,453 ± 588
	–160 cm	Wood	LU-147	11,830 ± 110	13,748 ± 167
	–225 cm	Mammoth tusk	LU-149	12,240 ± 160	14,352 ± 348
		Soft tissue	MAG-114	13,700 ± 80	17,085 ± 176
		Wood	MAG-117	11,870 ± 60	13,773 ± 144
		Wood	MAG-119	10,440 ± 100	12,351 ± 206
		Wood	MAG-118	10,260 ± 155	12,026 ± 355
Trail Creek Cave 9	Lower clay	Bison bone	K-1327	13,070 ± 280	16,030 ± 640
	Lower clay	Horse bone	K-1210	15,750 ± 350	19,010 ± 320
Ester Creek	Muck	Mammoth bone	CAMS-13301	11,310 ± 120	13,218 ± 148
	Muck	Mammoth bone	CAMS-13303	12,220 ± 70	14,275 ± 248
	Muck	Mammoth bone	CAMS-17126	12,270 ± 70	14,342 ± 267
	Muck	Mammoth bone	CAMS-17127	12,380 ± 70	14,548 ± 317
Fairbanks Creek	Muck	Mammoth carcass	SI-453	15,380 ± 300	18,629 ± 363
	Muck	Mammoth hair	CAMS-19253	17,170 ± 70	20,548 ± 398
	Muck	Mammoth carcass	CAMS-19250	17,550 ± 80	21,013 ± 386
Broken Mammoth	CZ 4[c]	Charcoal	UGA-6257D	11,040 ± 260	12,991 ± 234
	CZ 4	Charcoal	WSU-4265	11,280 ± 190	13,186 ± 213
	CZ 4	Hearth charcoal	CAMS-5358	11,420 ± 70	13,322 ± 142
	CZ 4	Swan bone	CAMS-8261	11,500 ± 80	13,400 ± 130
	CZ 4	Hearth charcoal	WSU-4262	11,510 ± 120	13,405 ± 150
	CZ 4	Ivory collagen	AA-17601	11,540 ± 140	13,439 ± 166
	CZ 4	Charcoal	WSU-4351	11,770 ± 210	13,707 ± 277
	CZ 4	Charcoal	WSU-4364	11,770 ± 220	13,716 ± 292
	CZ 4	Mammoth tusk	CAMS-9898	15,830 ± 70	19,084 ± 215
Mead	CZ 4	Charcoal	Beta-59121	11,560 ± 80	13,448 ± 122
	CZ 4	Charcoal	Beta-59120	11,600 ± 60	13,480 ± 118
	CZ 4	Mammoth ivory	CAMS-17408	17,370 ± 90	20,915 ± 411

(Table 4.1 continued)

Site	Level[a]	Material Dated	Lab No.	Uncalibrated Date	Calibrated Age[b]
Swan Point	CZ 4	Bone collagen		11,369 ± 50	13,264 ± 123
	CZ 4	Wood charcoal	CAMS-4252	11,660 ± 70	13,547 ± 141
	CZ 4	Wood charcoal	CAMS-12389	11,660 ± 60	13,546 ± 137
	CZ 4	Artifact residue	AA-19322	11,770 ± 140	13,672 ± 201
	CZ 4	Charcoal	Beta-QA-619	12,040 ± 40	13,910 ± 80
	CZ 4	Ivory collagen	CAMS-17045	12,060 ± 70	14,010 ± 160
	CZ 4	Carbonized grease	Beta-209883	12,100 ± 40	14,070 ± 120
	CZ 4	Carbonized grease	Beta-175491	12,110 ± 50	14,140 ± 247
	CZ 4	Carbonized grease	Beta-209884	12,220 ± 40	14,250 ± 170
	CZ 4	Charcoal	Beta-209882	12,290 ± 40	14,330 ± 160
	CZ 4	Carbonized grease	Beta-170457	12,360 ± 60	14,450 ± 160
Lime Hills Cave 1	114–115 cm	Bone collagen	Beta-67671	13,130 ± 180	16,138 ± 579
	91–96 cm	Bone collagen	Beta-67669	15,690 ± 140	18,979 ± 249
	70–77 cm	Bone collagen	Beta-67670	27,950 ± 560	32,647 ± 894
Bluefish Caves 1		Moose	CAMS-23472	11,570 ± 60	13,466 ± 80
		Horse	GSSC-2881	12,900 ± 100	15,758 ± 515
		Horse	RIDDL-278	17,440 ± 220	20,952 ± 434
		Mammoth	CRNL-1220	12,845 ± 250	15,670 ± 710
		Mammoth	RIDDL-559	13,940 ± 160	17,333 ± 167
		Mammoth	GSC-3053	15,540 ± 130	18,838 ± 208
		Mammoth	RIDDL-330	17,880 ± 330	21,369 ± 581
Bluefish Caves 2		Bison	RIDDL-561	10,230 ± 140	11,978 ± 340
Bluefish Caves 3		Bison	CRNL-1236	12,370 ± 440	14,850 ± 858
		Bison	RIDDL-279	13,390 ± 180	16,438 ± 577

SOURCES: Mochanov 1977:77–79; Goebel and Slobodin 1999:108–110, table 1; Larsen 1968:58–62; Dixon 1999:54–57; Holmes 2001:158, fig. 2; C. E. Holmes, pers. comm., 2006; Ackerman 1996a:473, table 10-6; Guthrie 2006

[a] Blank space indicates data not available

[b] Dates calibrated with the Cologne Radiocarbon Calibration Programme CalPal-SFCP-2005 glacial calibration curve (September 2005) (www.calpal.de)

[c] CZ, Cultural Zone

stream action (Vereshchagin 1977:42–49; Vereshchagin and Ukraintseva 1985:112) in a manner similar to the accumulations found in central Alaska (e.g., Péwé 1975:98). Alternatively, the high proportion of prime-age adults suggests "catastrophic mortality" and the possibility that most of the mammoth remains were derived from the destruction of a group (Vereshchagin 1977:46).[2]

While investigating the "cemetery" in 1970, Vereshchagin (1977:40) recovered several artifacts—along with bones of hare and wolf—on the bank roughly 120 meters downstream of the mammoth bones. These were reported to Mochanov (1977:77), who investigated the archaeology of the site in 1971–1973 and 1981 (Vereschagin and Mochanov 1972; Mochanov and Fedoseeva 1996c:218). Mochanov (1977:79) initially recovered 127 stone and 49 bone and ivory artifacts from the sediment and collected another 60 artifacts from the eroding bank. A few additional surface finds were reported from later visits (Vereshchagin and Ukraintseva 1985:107–109; Goebel and Slobodin 1999:121).

The artifacts in stratigraphic context were recovered from thin lenses of sand (1–5 millimeters thick) contained in a thick unit of bedded sandy loam. Both artifacts and bones were found in lenses at depths of 250 and 252–253 centimeters below the modern surface. In places, the two lenses converged, and they apparently represent a single occupation layer (Mochanov 1977:78–79). The sandy loam lenses were filled with fragments of willow and larch (Mochanov and Fedoseeva 1996c:218).

Neither the stone nor the nonstone artifacts recovered from the sediment at Berelekh are especially diagnostic (e.g., Slobodin 2001:38), and archaeologists have struggled to characterize the assemblage. A wedge-shaped microblade core and bifacial point were subsequently recovered at the site (Mochanov et al. 1991:214, fig. 140), but their relation to the other artifacts is unclear.[3] In 1980, Vereshchagin found a teardrop-shaped point that resembles the "Chindadn points" of eastern Beringia (see chapter 5) on the eroding bank (Vereshchagin and Ukraintseva 1985:109, fig. 5) (Figure 4.4).[4] Stone artifacts recovered from the sediment include two large biface fragments, one multifaceted burin, and several retouched blade-like flakes (Mochanov 1977:79–83; Abramova 1989:233–234). Roughly fifty fragments of bone and ivory are classified as artifacts, but none of them appears to be a formal tool.

Figure 4.4 Small bifacial point recovered from Berelekh by N. K. Vereshchagin in 1980 (photograph by JFH courtesy of N. K. Vereshchagin 1991); although similar to points found in central Alaska during the Lateglacial, this artifact was found out of context and cannot be dated

The most interesting artifacts from Berelekh are the ornaments and art, which are highly unusual for Beringia. Four pendants in the form of small flat pebbles were found in the sediment—each pebble wholly or partially perforated with a rotary drill from both sides (Mochanov 1977:80–82). A fifth pendant was recovered from the bank. One of the pendants exhibited a series of small incisions along its edge. And in 1965, an ivory fragment engraved with the image of a mammoth was obtained from a local inhabitant, who apparently had extracted it from a riverbank exposure located 50 kilometers upstream of the site (Bader and Flint 1977; Mochanov 1977:83–85). The engraving may or may not be connected with the occupation at Berelekh.

In addition to the mammoth "cemetery," more than 1,000 identified bones were found in association with the artifacts (table 4.2). At least two bones of fish also were recovered with the artifacts (Mochanov 1977:79).[5] In contrast to the bones and tusks from the mammoth "cemetery," these remains are assumed to represent mammals and birds either collected or consumed (or both) by the occupants of the site (Mochanov 1977:79), although this assumption is based on stratigraphic association; no occupation floor patterns could be mapped at the site. Another 178 burned and fractured mammoth bones were found on the eroding bank and also thought to be associated with the occupation (Mochanov and Fedoseeva 1996c:221).

Table 4.2

Vertebrate Remains from Berelekh

	No. of Specimens	Percentage of Specimens
Canis lupus (Wolf)	18	2
Lepus tanaiticus (Hare)	850	81
Mammuthus primigenius (Mammoth)	79	8
Equus lenensis (Horse)	1	<1
Rangifer tarandus (Reindeer)	2	<1
Lagopus lagopus (Ptarmigan)	92	9
Anser fabalis (Goose)	4	<1
Total	1,046	100

SOURCE: Vereshchagin 1977:42, table 12

Both the dating and interpretation of Berelekh are controversial (e.g., Goebel and Slobodin 1999:120–122; Pitul'ko 2001:267). Some of the uncertainty surrounding the site is related to the severe postdepositional frost disturbance, slumping of the exposed bank, and extreme difficulty of excavating the archaeological layer in frozen and disturbed sediment (Mochanov 1977:86–87). The artifacts may lie in a secondary or redeposited context (Slobodin 2001:38). Furthermore, Mochanov believes that much of the archaeological site was destroyed by the use of pressure hoses to collect the bones from the mammoth "cemetery" in 1970–1971 (Mochanov and Fedoseeva 1996c:218).[6]

Although only a small number of artifacts were recovered from Berelekh, this may be due to loss of materials from the use of pressure hoses at the site by the paleontologists. The presence of ornaments and art is not consistent with a small and short-term occupation but is more typical of a large residential camp. In fact, Mochanov (1977:86) suggested that the mass of mammoth bones might represent the remains of dwelling structures similar to those of Mezhirich and other late Upper Paleolithic sites on the central East European Plain.[7] These sites also date to the end of the LGM and beginning of the Lateglacial period, and they may have been located near large natural concentrations of bone (Soffer 1985; Hoffecker 2002a:232).

The burned mammoth bone collected from the eroding bank suggests that the mammoth "cemetery" may have been mined as a source of fuel in a landscape containing many shrubs but few if any trees. The recovery of soft tissues indicates that at least some of the remains were frozen in a fresh or "green" state and therefore flammable, while the woody shrub fragments show that starter fuel was available (box 4.1, figure 4.5). Disturbance to the occupation layer, however, precluded identification of any former hearths or pits containing bone ash (i.e., more direct evidence of bone fuel use found in many central East European Plain sites [Hoffecker 2002a:200–212]).

The presence of the mammoth "cemetery" and associated human occupation may explain why the earliest known site in western Beringia is located in the far north. It is in the large lowland basins of the north where substantial quantities of mammal bone would have accumulated at places like Berelekh during the LGM. And it is across the lowland areas that woody shrubs first began to spread during the early Lateglacial.[8]

In sum, Berelekh probably represents the earliest evidence of settlement in western Beringia during the Lateglacial, although some uncertainty still surrounds the dating of the site (new radiocarbon dates on bone obtained from the occupation level in 2004 yielded ages of 14.2 and 13.3 cal ka [Pavlova et al. 2006]). And its occupants probably made artifacts similar to those of the Dyuktai culture (including wedge-shaped microblade cores), although here, too, there is some ambiguity. Perhaps the most interesting information that Berelekh offers about Lateglacial settlement in Beringia is the apparent importance of small mammals—especially hare—birds, and possibly fish in the economy. Assuming that the remains found with the artifacts represent food refuse, they indicate that, from the outset, the Beringians were subsisting on a broad-based diet of northern interior resources (Hamilton and Goebel 1999:184).

Eastern Beringia

One of the striking characteristics of the archaeological record of early settlement in Beringia is that, with the exception of the Yana RHS locality, no significant discoveries have been reported from its western half for more than 30 years. In fact, with the redating of the lowest horizon at Ushki, there has been some attrition of remains in western Beringia thought to represent the earliest post-LGM occupation. This is not the case in eastern Beringia, however, where several major finds during the past 15 years have significantly altered the picture of early settlement. Especially important are a group of small sites on the

Box 4.1

The Problem of Fuel in Beringia

Fuel may have been the critical variable in the Lateglacial occupation of Beringia. It appears unlikely that either lack of food sources or severe climates were settlement barriers between 20 and 15 cal ka (Guthrie 1990:273–279). Significantly, permanent settlement at 15–14 cal ka coincides with a major expansion of woody vegetation in the form of shrubs like willow and dwarf birch ("Birch Zone"). Moreover, an earlier intrusion into the Arctic and northwestern Beringia during a warmer interval at 30 cal ka also took place when woody vegetation was locally available (Anderson and Lozhkin 2001).

The problem of fuel in Beringia is part of a wider issue concerning the role of controlled fire in the colonization of higher latitudes. Controlled fire recently has been documented in the Near East at 0.79 mya (Goren-Inbar et al. 2004) and was used heavily by the Neandertals throughout their geographic range (Mellars 1996:295–301). Before modern humans, however, fire-making technology probably was not present (some recent foragers in lower latitudes lacked it) and, with occasional exceptions, fuel was confined to wood (Villa, Bon, and Castel 2002).

After 50 cal ka, modern humans expanded into areas of northern Eurasia where trees were, or subsequently became, scarce or absent. In these areas, they often turned to alternative fuels in the form of fresh bone, lignite (or combustible shale), and probably dung (e.g., Goebel 1999:218; Soffer 2000:60; Hoffecker 2002b:125). Dung fuel has lower archaeological visibility than bone or lignite but may be detected in concentrations of plant phytoliths (e.g., Guthrie 1983a:282–283). Modern humans developed a variety of fire-related technologies, including—probably—fire-making drills, interior constructed hearths within artificial shelters, and portable lamps that burned animal fat (de Beaune and White 1993).

The use of alternative fuels presented special problems that may explain when and how modern humans occupied Beringia. Experimental research reveals that fresh bone and lignite have comparatively high ignition temperatures of 350–380°C and 500°C, respectively (Théry-Parisot 2001:136). According to these experiments, wood tinder is necessary to ignite bone and must make up at least 20 percent of the fuel (Théry-Parisot 2001:140). Although recent experiments suggest that under some circumstances it may be possible to ignite bone without wood, the results are preliminary (Barbara Crass, pers. comm., 2005). Another requirement of bone is that it must be kept fresh or "green" in order to produce heat.

Large-mammal dung also offers a potential fuel source, and there are many examples of its use in late prehistoric and ethnographic contexts—typically

(Box 4.1 continued)

in arid steppe or grassland areas (e.g., Miller 1984; Anderson and Ertug-Yaras 1998). Experimental research indicates that dung fires yield less heat than wood or fresh bone (Wright 1986), and ethnographic observations suggest that large quantities of dung must be gathered daily to provide adequate fuel (Rhode, Madsen, and Brantingham 2003).

The use of lignite or combustible shale during the Pleistocene appears relatively rare, which probably reflects its scarcity in surface exposures, as well as its high ignition temperature. The use of bone, in contrast, is widely reported on the East European Plain before and after the Last Glacial Maximum (LGM). Fresh bone may have been gathered from natural concentrations at the mouths of tributary streams—where sites often were located. At the sites, large quantities of unburned bone have been found in deep pits that may have been dug to the base of the thaw layer and were used like "ice cellars" to keep bone fuel fresh during warmer months (Vereshchagin and Baryshnikov 1984:492–493; Hoffecker 2002a:226) (figure 4.5).

Between 20 and 15 cal ka, Beringia contained a small number of trees (poplar) and shrubs (willow), isolated sources of exposed lignite (e.g., Nenana Valley, central Alaska), and abundant supplies of bone and dung. Human inhabitants could have obtained fresh bone from kills and also from concentrations of partially frozen bone in stream valleys of major lowland basins (Péwé 1975:98), but it would have been difficult to ignite in the absence of wood tinder. Dung could have been ignited without wood tinder, and it has been suggested that it would have provided an adequate fuel source for humans at this time (Rhode, Madsen, and Brantingham 2003).

During 15–14 cal ka, shrub tundra spread across the Beringian lowlands and altered the landscape with respect to fuel sources. Although trees remained scarce, woody shrubs increased significantly; willow may have been the most important as fuel. The supply of fresh bone and dung probably decreased somewhat as mammoth, horse, and other large grazers of the steppe-tundra declined. Human occupations in Beringia dating to this interval and the succeeding few millennia (14–10 cal ka) contain much evidence for burning bone and some wood charcoal—identified occasionally as willow or poplar (Holmes, VanderHoek, and Dilley 1996:321, table 6-5; Pearson 1999:339). Ethnohistoric accounts (e.g., Gubser 1965) suggest that, where they were locally abundant, willow shrubs by themselves might have provided sufficient fuel. At most sites, however, bone and wood were burned together, and the latter probably was used to ignite the former. In upland areas, natural bone accumulations probably were rare, and there is some evidence for the use of dung and lignite (see chapter 5).

Figure 4.5 Pit containing bone at Kostenki 11, probably dating after 20 cal ka and reflecting heavy use of bone fuel on the central East European Plain (photograph by JFH)

Tanana River in central Alaska that provide the most reliably dated traces of occupation in Beringia before 13.5 cal ka (Holmes 2001).

TRAIL CREEK CAVES

In 1948, David M. Hopkins investigated a series of limestone caves located on a small creek near Imuruk Lake in the northeast Seward Peninsula. During 1949–1950, the Danish archaeologist Helge Larsen excavated Trail Creek Cave 2 and Cave 9, recovering historic and prehistoric artifacts and faunal remains from various periods (Larsen 1968:13–16). Jean Schaaf undertook new excavations at five caves in 1985 (Schaaf 1988).

In a deeply buried layer of sandy silt and rubble in Cave 2, Larsen (1968:71–72) found microblades and slotted antler points associated with a caribou bone that dated to slightly more than 10 cal ka. The points remain rare examples of weapons slotted for microblade insets in Beringia, similar to those recovered from southern Siberia (see chapter 3). Near the entrance to Cave 9, Larsen encountered isolated horse and bison bone fragments in the basal clay layer that yielded calibrated dates of 19 and 16 cal ka, respectively. Comparison with specimens from other sites convinced Larsen that the bison fragment had been modified by humans, and the recovery of dog teeth seemed to provide further evidence of human occupation in the lowest layer (Larsen 1968:58–62).[9]

For many years, the faunal remains from the lowest layer of Trail Creek Cave 9 were considered the oldest traces of humans in Alaska (e.g., West 1981:175). And they established a precedent for the identification of a human presence in Beringia on the basis of modified bones (e.g., Irving 1978). In recent years, however, the "dog" teeth were reclassified as deciduous bear teeth, while taphonomic studies demonstrated that the bison bone probably was gnawed by carnivores (Dixon and Smith 1986; Vinson 1993; Hamilton and Goebel 1999:179). The discoveries from the lowest level at Trail Creek Caves are described here because of their impact on research in Beringia during the 1970s and 1980s.

FAIRBANKS MUCK DEPOSITS

The large concentrations of mammal bone that accumulated at places like Berelekh are especially common in the lowland basins of central Alaska. During the LGM, bones were concentrated at stream confluences and buried in frozen silt (often referred to as "muck") derived primarily from loess (Péwé 1975:95–101). Many of these bones have been washed out of the muck by placer miners using high-pressure hoses to thaw the frozen sediment and access gold-bearing stream deposits (Guthrie 1990:65–68). At least some of the bones are fresh and attached to pieces of hide, hair, and musculature. As in Siberia, frozen carcasses occasionally are recovered—such as the nearly complete bison carcass named "Blue Babe" in 1979 (Guthrie 1990).

Late Pleistocene muck deposits are common in the Fairbanks area, and extensive placer mining of these deposits has yielded thousands of large mammal bones of late Pleistocene age.[10] The predominance of steppe bison, horse, and mammoth among them provided evidence of a grassland habitat in the area during the LGM (Guthrie 1968). Radiocarbon dates on the bones—many of them on the collagen fraction—and the soft tissues indicate that they accumulated during 40–13 cal ka (Péwé 1975:98–101; Matthews 1982:139–142; Guthrie 1990:240–245).

Figure 4.6 Bone point recovered from the Fairbanks muck at Goldstream Creek in 1933 and recently dated by AMS radiocarbon to less than 10 cal ka (from Rainey 1940:306, fig. 16; reproduced by permission of the Society for American Archaeology from *American Antiquity,* vol. 5, no. 4, 1940)

The Fairbanks muck deposits also yielded a small number of artifacts, and these were recognized as possible evidence of human settlement in eastern Beringia at this time (Rainey 1940:304–307). They included two bone points recovered from Goldstream Creek in 1933 (figure 4.6), which recently were dated directly by AMS (accelerator mass spectrometry) radiocarbon method to slightly less than 10 cal ka (Dixon 1999: 52–53). Two bifacial stone points—at least one of which was reportedly associated with mastodon jaw fragments—were found in muck deposits at a depth of 18 meters on Ester Creek in 1939 (Rainey 1940:305). The jaw fragments were recently dated to about 14 cal ka (Dixon 1999:54).

During the 1940s, two more stone artifacts were found in possible association with frozen mammoth parts on Fairbanks Creek. The artifacts include a chert biface and scraper. The biface was recovered in 1940 near a mammoth foot that was later dated to 21–18.6 cal ka; the scraper was found in 1948 near the partial carcass of a juvenile mammoth that was subsequently dated to more than 25 cal ka. In both cases, the association of the artifacts with the mammoth remains is problematic (Dixon 1999:55–57).

The archaeology of the Fairbanks muck deposits suffers from the same problems that plague the Berelekh site. The conditions of deposition (i.e., fluvial processes), frost disturbance, and the use of pressure hoses to thaw the sediment make it difficult to establish the provenience and association of artifacts. Only direct dating of artifacts—such as the bone points—provides a means of determining their age. The muck deposits have yet to produce firm evidence of settlement before 10 cal ka.

TANANA RIVER SITES

The most concrete evidence of human settlement in Beringia before 13.5 cal ka is found in three open-air sites located along the Tanana

River in east-central Alaska. Only these sites—discovered and excavated during the past 15 years—provide unequivocal traces of occupation in this time range. The artifacts are buried in a relatively undisturbed stratigraphic context, and their chronology is supported by a series of radiocarbon dates and cross-correlation of strata. Moreover, the Tanana River sites have yielded a wealth of associated faunal remains that offer valuable insights to the ecology of the Beringians during this early phase of settlement.

The three sites are located on the north side of the Tanana River roughly 100 kilometers southeast of Fairbanks (latitude 64°N). The Broken Mammoth site—near the mouth of Shaw Creek—was discovered in 1989 by Charles E. Holmes (1996:312), who has researched early prehistoric sites in interior Alaska for more than three decades (figure 4.7). In the early 1990s, Holmes also found artifacts at the Mead site, which is approximately 1 kilometer north of Broken Mammoth and had been investigated earlier by geologists. Swan Point was discovered in 1991 on

Figure 4.7 Excavation at Broken Mammoth in 1991, where—along with two other nearby sites in the Tanana Valley—the earliest firmly dated occupation in Beringia was discovered in a deep sequence of eolian sediment overlying a bedrock bluff (photograph courtesy of Charles E. Holmes)

the northern edge of Shaw Creek Flats, roughly 7 kilometers northeast of Broken Mammoth (Holmes, VanderHoek, and Dilley 1996:319).

All three sites occupy low bedrock surfaces capped with eolian sand and silt (loess) derived from the Tanana River floodplain. The eolian deposits at Broken Mammoth and Mead are slightly less than 2 meters in depth, while at Swan Point they are less than 1 meter in thickness (figure 4.8). At all three sites, the lowest occupation level rests in loess near the base of the eolian sequence and is associated with a buried soil complex (Holmes 1996:313–314; 2001:158, fig. 2). The buried soil complex and the occupation level are dated to 14.5–13 cal ka by a total of fourteen conventional and AMS radiocarbon dates (table 4.1). The age of the lowest level also is supported by dates on the overlying sediments and occupation levels.

Few features have been reported from the lowest occupation level at the three sites—no remains or traces of former dwellings or pits—but former hearths were excavated at Broken Mammoth and Swan Point (Holmes 1996:317). At Swan Point, analysis of burned residue revealed use of bone fuel (Holmes and Crass 2003), and wood charcoal was identified as willow (*Salix* sp.) and willow/poplar (*Salix*/*Populus*) (Holmes, VanderHoek, and Dilley 1996:321, table 6-5).

Although most of the artifacts recovered from the lowest occupation level at each site represent flakes and other debris from the manufacture and sharpening of stone tools, recent excavations at Swan Point have yielded an assemblage containing more diagnostic artifacts. In 2003, a wedge-shaped microblade core and platform preparation flake were found (figure 4.9), and since then five more microblade cores have been discovered in this level. Other artifacts include additional core preparation flakes (in the form of ridge spalls, ski spalls, and core tablets) and more than 320 microblades. Burins include both dihedral and transverse forms, and burin spalls are present (Holmes, VanderHoek, and Dilley 1996:321; Holmes 1998, 2001; C. E. Holmes, pers. comm. 2006) (Figure 4.10).

The artifacts from Broken Mammoth are less diagnostic and include several scraping tools and some bifacial thinning flakes (Holmes 1996:317), while only fragments of a biface and scraper have been reported from Mead (Hamilton and Goebel 1999:167–168).

The wedge-shaped microblade cores and associated core preparation pieces from Swan Point exhibit a core technique (i.e., Yubetsu [Morlan

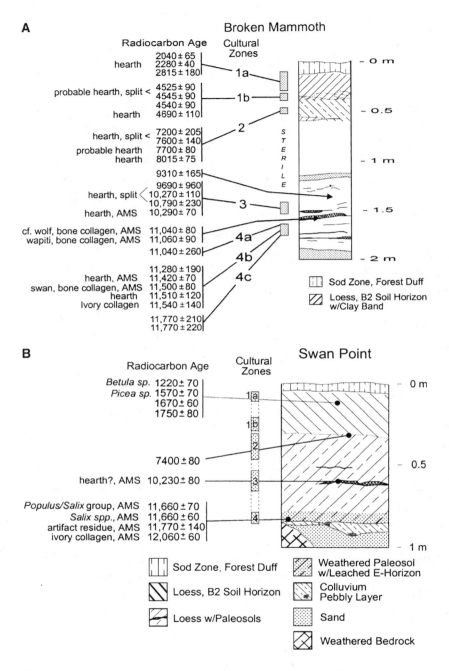

Figure 4.8 Stratigraphy and uncalibrated radiocarbon dates of Broken Mammoth and Swan Point Tanana Valley sites (after Holmes 2001:158, fig. 2; *Arctic Anthropology Journal,* vol. 38, no. 2. Copyright 2001. Reprinted by permission of University of Wisconsin Press)

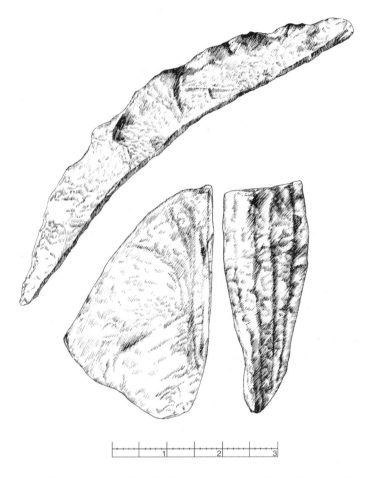

Figure 4.9 Wedge-shaped microblade core recovered from Cultural Zone 4 at Swan Point in 2003 (drawn from a photograph provided by Charles E. Holmes)

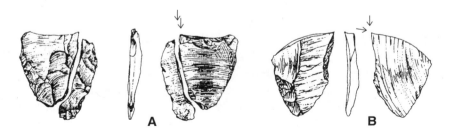

Figure 4.10 Artifacts from the lowest occupation level at Swan Point (courtesy of Charles E. Holmes)

1978a]) found in Northeast Asian sites, including those of the Dyuktai culture (C. E. Holmes, pers. comm., 2006). Dyuktai assemblages also contain transverse and dihedral burins (e.g., Mochanov and Fedoseeva 1996a:167–174). The earliest firmly dated post-LGM horizon in Beringia is therefore linked to Northeast Asia and specifically to the Dyuktai culture of the Lena Basin (Holmes and Crass 2003) (box 4.2, figure 4.11).

Obsidian imported from distant sources was found at Broken Mammoth and Swan Point. At each site, volcanic glass from both Batza Téna in northwest Alaska (Koyukuk River drainage) and the Wrangell Mountains in southeast Alaska was identified (Holmes 2001:167). Both sources are located hundreds of kilometers from Shaw Creek, and the appearance of materials imported from them indicates either large foraging ranges or trading networks, or both.

Bone and ivory are relatively well preserved at the Tanana River sites due to the high calcium carbonate content of the loess (Holmes 2001:154). At Broken Mammoth, two ivory points were found, along with another ivory implement (handle?) and several fragments of worked tusk—one of which contained an imbedded stone chip (Holmes 1996:317–318). Tusk fragments were also found at Swan Point, and an ivory point tip is reported from Mead (Hamilton and Goebel 1999:167). Some of the ivory pieces yielded radiocarbon dates older than the mean for the lowest occupation level (including a worked tusk fragment from Broken Mammoth dated to 19 cal ka), suggesting that the occupants may have been scavenging mammoth ivory from older contexts.

Box 4.2

Microblade Technology in Beringia (by Craig M. Lee)

Microblades, and the formal, polyhedral cores on which they were made, are hallmarks of Beringian technology. Nels Nelson (1935, 1937) was the first to recognize that similarities among cores found in Alaska and Asia indicated a connection between the two regions, although whether that connection represents the migration of people or the diffusion of technology remains unresolved. The late Pleistocene/early Holocene microblade technology of eastern Beringia is described as belonging to the American Paleoarctic tradition (APAt), as defined by Douglas Anderson (1968, 1970,

(Box 4.2 continued)

1988), with many discoveries in central Alaska being ascribed to the Denali complex, as defined by F. H. West (1967). In western Beringia, early microblade technology is assigned to the Dyuktai culture (Mochanov 1977) and in later periods to Sumnagin (Slobodin 1999).

Characteristics of the microblade cores in Beringia include (1) sharp-edged keels or bases, which are frequently bifacially flaked; (2) blade removal from only one face, that is, frontally faced; and (3) a generalized wedge-shaped morphology when viewed from above or head-on toward the flutes (West 1967). While the ultimate origin of blade and bladelet technology may lie in the emergent Upper Paleolithic technology of the Levant some 50 cal ka (Meignen and Bar-Yosef 2002), the blade and bladelet technology of northern Eurasia is the likely progenitor of the formal mode of microblade production described above, which first appeared in the Transbaikal region about 20 cal ka (Goebel et al. 2000).

Microblade production requires that flakes be driven entirely off the face of the core to allow for subsequent blade removals. Consequently, high-quality materials (e.g., chert and obsidian) without inclusions or natural checks that could derail production were preferred. Although experimentation has shown that microblades can be produced in a variety of ways, their morphological regularity and the small size of most cores suggest that a vise to support the core against an anvil was required and that pressure flaking was probably used to detach the microblades (e.g., Morlan 1970, 1978a; Flenniken 1987; Tabarev 1997).

Unlike bifacial lithic tool production, which involves tool life-cycles or stages of modification and reuse, microblades are essentially in their finished form when they are removed from the core. Most if not all of their utility derives from how they are hafted (i.e., attached to a handle or shaft). Because of their small size, most archaeologists believe that microblades were used as insets in a haft that was carved or ground out of available organic materials such as bone, antler, and ivory.

The first projectile points slotted to receive microblades discovered in North America were found in Alaska in 1950 at Trail Creek Cave 2 on the Seward Peninsula. The Trail Creek Cave points are oval in cross-section with diametrically opposed grooves running from the tip to the mid-point of a beveled butt, which was frequently roughened to facilitate lashing (Larsen 1968). Following Medvedev (1964), who was familiar with Siberian discoveries, and MacNeish (1954) and Anderson (1968), who were working in North America, Helge Larsen (1968) described the slotted projectile points recovered at the Trail Creek Caves as arrowheads. The recent discovery of a

(Box 4.2 continued)

slightly younger (ca. 8 cal ka) slotted antler projectile point and dart shaft in an ice patch in the Yukon Territory suggests, however, that some weapons armed with slotted points may have been propelled by atlatls (Hare et al. 2004).

Although slotted projectile points have primarily been recovered from interior sites, examples are also known from coastal localities. The artifact assemblage from the early/middle Holocene Rice Ridge site on Kodiak Island contains at least three slotted points and two possible preforms (Steffian, Eufemio, and Saltonstall 2002). Some of the slotted points are made from marine mammal bone, whereas others are made of antler. A cluster of microblades recovered from a constricted passage at On-Your-Knees Cave in the Alexander Archipelago of southeastern Alaska are also believed to represent the remains of a slotted projectile point, although the organic component is absent (Lee and Dixon 2006). Other projectile points slotted to receive microblades have been found in mid-Holocene components at sites further south on the northwestern coast; these include an antler point at the Cohoe Creek site (Christensen and Stafford 1999, 2005) and a small fragment made of marine mammal bone at Namu (Carlson 1996).

Slotted projectile points have been discovered at sites in Northeast Asia, including Zhokhov Island (Pitul'ko and Kasparov 1996:8–9, figs. 5 and 6), Chernoozer'e II (Gening and Petrin 1985:48, fig. 17), and Kokorevo I and Studenoe I (Vasil'ev 2001:14, 21, figs. 7 and 12). Several of these points still retain inset microblades (figure 4.11).

There is evidence to suggest the projectile points found on Zhokov Island may have been used for hunting polar bear. A more common use, however, probably was for taking terrestrial game such as caribou and bison—the latter supported by the recovery of a bison scapula pierced by a slotted projectile point at Kokorevo (Abramova 1979:29, fig. 16). The absence of points retaining the actual inset microblades in eastern Beringia is undoubtedly due to preservation bias rather than technological difference. Discoveries at younger sites with exceptional preservation in areas proximate to the former Beringian landmass (e.g., the northwestern coast of North America) demonstrate that some microblades were used in end-hafted knives. Such tools would have been ideal for the production of fine craft work such as the manufacture of clothing or for specialized processing tasks such as cleaning fish (Croes 1995; Bowers et al. 1996; Barclay, Malmin, and Croes 2005).

The form and construction of slotted points in eastern Beringia appears to be morphologically variable. Examples from the Ilnuk and Lime Hills sites in the Kuskokwim drainage of southwestern Alaska are unilaterally and bilaterally grooved, respectively, with the latter having cut—as

(Box 4.2 continued)

opposed to gouged—grooves (see chapter 6). Although some microblades may have been held in place primarily by the pressure imposed by the haft (e.g., Guthrie 1983b), the rounded bottoms of the grooved points and the relatively thin side walls suggest that friction alone was not enough to hold the microblades in place. It is more likely that gum-like mastic that could be cured and reversed was used to bind the microblades in place and to allow for the cutting edge to be made regular and maintained. Traces of possible mastic have been found on the slotted projectile point mentioned previously from the Yukon Territory (Hare et al. 2004; G. Hare, pers. comm., 2005).

Faunal remains were especially well preserved at Broken Mammoth, and individual bones and teeth from the lowest occupation level represent a variety of mammals and birds (table 4.3). With the exception of the large quantity of ground squirrel bones, many of which were found as intact skeletons in fossil burrows or krotovinas (Yesner 2001:321), the majority of remains probably were brought to the Broken Mammoth site by its occupants and represent food debris. Elk (or wapiti) is the most common large mammal; hare is equally common. It is worth noting that besides the mammoth tusk fragments—possibly scavenged from natural bone accumulations—the fauna is essentially postglacial or modern. Waterfowl are dominant, suggesting a later winter/spring occupation (Yesner 2001:322).

Only preliminary faunal data are available for the lowest level at Swan Point, which also contains waterfowl, including goose (*Branta* sp.), mammoth, and large cervid remains (Holmes, VanderHoek, and Dilley 1966:321; Holmes and Crass 2003). Remains of wapiti and goose are reported from Mead (C. E. Holmes, pers. comm., 2006).

The Tanana River sites provide a remarkable glimpse of life in Beringia before 13.5 cal ka. Analysis of pollen cores from nearby lakes reveals that these sites were first occupied shortly after a shrub tundra environment became established in the Tanana Basin (Bigelow and Edwards 2001; Bigelow and Powers 2001:181–182). Even more so than at Berelekh, the faunal assemblages indicate a broad-based diet of northern interior resources with particular emphasis on small mammals and birds. One of the most striking features of the Broken Mammoth site is that similar faunal remains are found throughout }the sequence of occupations, which extends into the late Holocene

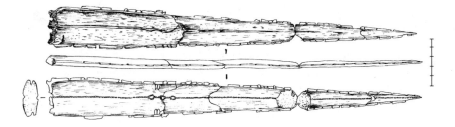

Figure 4.11 Slotted projectile point with inset microblades from Chernoozer'e II in western Siberia (after Gening and Petrin 1985:48, fig. 17)

(Holmes 1996:314), and the same pattern was observed by Cook (1969) at Healy Lake (see chapter 5).

LIME HILLS CAVE 1

Another possible human occupation before 13.5 cal ka is reported from southwest Alaska. It is found in the Lime Hills area, which adjoins the northern flank of the Alaska Range and is drained by the Stony River—a major tributary of the Kuskokwim. In 1992, a geologist recognized the archaeological potential of a small cave on the eastern margin of the Lime Hills. Robert Ackerman, who had been researching early prehistoric sites in southwest Alaska for some years, was urged to investigate the cave, which he did during the following year (Ackerman 1996a:470).

The research at Lime Hills Cave 1 bears many similarities to the earlier work at Trail Creek Caves on the Seward Peninsula (described above). The cave is situated at an elevation of 527 meters above sea level. The entrance measures 6.4 meters in width and 2–2.5 meters in height, and the interior extends roughly 18 meters into the bedrock. The total depth of deposits filling the cave appears to be no more than about 1.2 meters (Ackerman 1996a:472, fig. 10-6).

Few artifacts were recovered during the 1993 investigations, but Ackerman (1996a:470–474) distinguished an upper component dating to 11–10 cal ka and possible traces of an earlier occupation dating to 19–16 cal ka.[11] The upper component contained a microblade fragment and a laterally slotted point fragment of bone or antler (Ackerman 1996a:472, fig. 10-7). Among the identified faunal remains were caribou, hare, sheep, porcupine, bear, arctic fox, some rodents, birds, and fish (Ackerman 1996a:474), although it is not clear how many of them are associated with the artifacts.

Table 4.3

Vertebrate Remains from Cultural Zone 4 of Broken Mammoth[a]

	No. of Specimens	Percentage of Specimens
Mammuthus primigenius (Mammoth)	+[b]	+
Bison priscus (Steppe bison)	21	5
Cervus elaphus (Elk)	44	10
Alces alces (Moose)	4	1
Ursus sp. (Bear)	1	<1
Alopex lagopus (Arctic fox)	18	4
Lutra canadensis (River otter)	6	1
Lepus sp. (Hare)	44	10
Marmota flavescens (Hoary marmot)	5	1
Spermophilus parryi (Arctic ground squirrel)	305	66
Sorex arcticus (Arctic shrew)	11	2
Total Identified Mammals	459	100
Cygnus columbianus (Tundra swan)	525	66
Branta canadensis (Canada goose)	22	3
Anser albifrons (White-fronted goose)	54	7
Chen hyperborea (Snow goose)	35	4
Anas platyrhynchos (Mallard)	24	3
Anas acuta (Pintail)	36	4
Anas strepera (Gadwall)	4	<1
Anas americana (Widgeon)	2	<1
Anas carolinensis (Green-winged teal)	22	3
Lagopus lagopus (Willow ptarmigan)	77	10
Total Identified Birds	801	100

SOURCES: Yesner 2001:321–322, table 2; D. R. Yesner, pers. comm., 2006
  [a] Additional remains from this horizon include unidentified medium/large mammals
  (*n* = 329), unidentified small mammals (*n* = 173), microtine rodents (*n* = 151), and unidentified
  birds (*n* = 363)
  [b] Present, but no quantitative information given

In any case, many probably represent remains of animals that either inhabited the cave or were brought there by other animals.

From lower depths in the cave sediment—near the base of the latter—Ackerman (1996a:472–474) recovered a caribou metatarsal that exhibited possible traces of working and polish and yielded a radiocarbon date of about 16 cal ka and a caribou humerus with possible tool cut marks dating to 18.9 cal ka. Also found was a bison bone (astragalus) that yielded a pre-LGM radiocarbon date of 32.6 cal ka; apparently, it had been imported to the cave from an older context (Ackerman 1996a:475).

NORTHERN YUKON TERRITORY

The 1968 report of an early human occupation at Trail Creek Cave 9 on the basis of modified bones seems to have established a precedent that had major consequences for Beringian archaeology during the two decades that followed. In 1966, C. R. Harington inaugurated a long-term program of paleontological research in northern Yukon. During the first year of the program, Harington recovered a fleshing tool carved from a caribou tibia in the Old Crow Basin (Morlan and Cinq-Mars 1982:357–359). Analysis of the tibia indicated that it had been worked when the bone was fresh. A radiocarbon date of 27,000 + 3000/–2000 ^{14}C years BP (32 cal ka) was later obtained on this artifact (Irving and Harington 1973) and was critical in launching a major effort to locate traces of late Pleistocene occupation in the northern Yukon (e.g., Bonnichsen 1978; Irving 1978; Morlan 1986).

The Old Crow Basin is the largest of three major lowland basins in the northern Yukon that lie between latitudes 67 and 69°N (i.e., above the Arctic Circle). The area represented the northeastern edge of Beringia during the LGM—as close as 100 kilometers to the margin of the Laurentide Ice Sheet at its maximum extent (Morlan 1978b:79). Although ice free, all three basins filled with massive glacier-dammed lakes—fed in part by meltwater—during the LGM. At other times, millions of bones of mammoth, horse, bison, and other large mammals accumulated in their stream systems as in the Kolyma Basin and the creek bottoms around Fairbanks (Matthews 1982:139–141; Morlan and Cinq-Mars 1982:359).[12]

In the years after the published date on the caribou flesher tool, hundreds of large mammal bones were recovered from the northern Yukon that were thought to represent artifacts or human food debris of the pre-LGM period.[13] Most of them were found at Old Crow Flats (Morlan 1978b:80). However, taphonomic studies of bone breakage—which became especially common after 1980—suggested that natural causes could account for the types of damage found on the bones from the northern Yukon (e.g., Haynes 1983a). And during the mid 1980s, the original caribou flesher tool was re-dated to the late Holocene (Nelson et al. 1986).

The intensive research program in the northern Yukon did yield an important archaeological site on the southern edge of the Bluefish Basin. Three small caves were discovered in 1975 at an elevation of about

600 meters on a limestone ridge in the foothills of the Keele Range and investigated by Jacques Cinq-Mars and colleagues during 1977–1981 (Cinq-Mars 1982:20; Morlan and Cinq-Mars 1982:366–368; Cinq-Mars and Morlan 1999). The stratigraphy inside and near the entrances of the Bluefish Caves is relatively shallow, but the sediments yielded a rich array of fauna and a modest quantity of artifacts.

Above a basal layer of weathered bedrock lies a unit of loess that contains some lenses of organic material. The loess apparently dates to the LGM and Lateglacial and produced remains of mammoth, bison, horse, caribou, sheep, and other large mammals (Cinq-Mars 1990: 16–18). A horse bone from the upper part of the loess in Cave I was dated to 15.8 cal ka, and a mammoth bone found in the lower part of the same unit in Cave II was dated to 18.8 cal ka (Morlan and Cinq-Mars 1982:368). Smaller vertebrates included rodents, birds, and fish. A layer of organic-rich rubble overlies the loess and is capped with the modern soil.

Cinq-Mars and his colleagues recovered small stone flakes throughout the loess unit but noted that they were most common in the zone where major vegetation change is indicated by the pollen data (Cinq-Mars 1990:21–22). In this zone, the pollen reflects a shift from herbaceous tundra to shrub tundra (Birch Zone). Among the small assemblage of retouched pieces are several wedge-shaped microblade cores and burins. Also found were a number of modified bones thought to represent cores and tools (Cinq-Mars 1990:20–24; Cinq-Mars and Morlan 1999).[14]

The precise dating of the occupation in Bluefish Caves is not clear because of the problematic association of the dated bones and artifacts. The association of the bulk of the artifacts with the transition to shrub tundra is therefore particularly significant. Elsewhere in eastern Beringia, this transition takes place at higher elevations at roughly 13 cal ka (Bigelow and Powers 2001). Occupation of the Bluefish Caves probably took place during the later phases of the Lateglacial and not before 13.5 cal ka.

Summary and Conclusions

The timing of the post-LGM settlement of Beringia is currently fixed at 14.4–13.5 cal ka, which is based primarily on the dating of the lowest occupation levels in the Tanana Basin. The sample of sites and dates is small, however, and future discoveries may change the current estimate. As Holmes (2001:167) notes, the presence of imported obsidian in the Tanana Basin sites indicates a thorough knowledge of the landscape and suggests greater time depth of settlement in eastern Beringia. To be sure, some of the radiocarbon dates associated with the occupation layer at Berelekh are older than 15 cal ka (Mochanov 1977:77), but significantly older dates from Trail Creek Caves, Lime Hills Cave 1, and Bluefish Caves appear doubtful because of the problematic association of the dated material (modified and unmodified mammal bone) and the occupation (Dixon 1999:50–61; Hamilton and Goebel 1999: 178–179; Yesner 2001:316).

The first permanent settlement of Beringia correlates well with the rise in birch pollen and the shift from steppe-tundra to mesic shrub tundra. In fact, the base of the Birch Zone seems to provide a suitable pollen-stratigraphic marker for this event, and in some places (e.g., Bluefish Caves) may offer a chronological yardstick for the archaeology. Not only is there a correlation between the spread of people and shrub tundra into the lowlands at 15–14 cal ka, but also—as described in chapter 5—there is a correlation between later movement of shrub tundra and human occupation into upland areas as well (e.g., Bigelow and Powers 2001). Because neither change in temperature nor the availability of food resources seems likely to have been the critical variable, we conclude—as others have already suggested—that an increase in wood fuel was the factor that tipped the balance toward settlement of Beringia (Hoffecker and Elias 2003). Specifically, we argue that the rapid spread of woody shrubs such as willow and dwarf birch provided a fuel source and—perhaps even more important—a means of reaching the high ignition temperature of green bone (Hoffecker and Elias 2005). Most Lateglacial sites in Beringia yield traces of burned bone.

Perhaps the most striking aspect of the early Lateglacial archaeological record is the evidence for a broad-based diet similar to that of Holocene peoples of the northern interior. The Lateglacial Beringians had developed a postglacial shrub tundra economy with an emphasis on small mammals and birds—especially waterfowl—and hunting of large mammals such as elk, moose, and sheep (Hamilton and Goebel 1999:184; Yesner 2001:318–323). Typical steppe-tundra taxa of the LGM such as horse and woolly rhinoceros were either very scarce or extinct, and mammoth may have been exploited chiefly as a source of (scavenged) ivory. Although steppe bison was hunted, this large grazer survived the shrub tundra transition in steppe refugia and became part of the Holocene landscape (Guthrie 1990:285–286). The Beringian economy was a continuation of trends in Northeast Asia following the LGM (Goebel 1999:218–219) that are manifest at Dyuktai Cave (Mochanov and Fedoseeva 1996a:167–174).

Archaeologists were slow to recognize the character of the early Beringian industry due to the small samples of artifacts and problems of dating artifacts in frost-disturbed contexts. It is now clear, however, that the Lateglacial Beringians brought a microblade technology with them from Northeast Asia. As noted in chapter 3, this technology is an efficient one that conserves precious high-quality stone. It seems to be especially well suited to mobile foragers who need to minimize the amount of carried stone (Goebel 2002). And it is tied to production of slotted points of bone or antler that are manufactured with burins.

5

The End of the Lateglacial Interstadial

Although the ice core record indicates that temperatures were slightly cooler during the latter phases of the Lateglacial 13,500–12,800 years ago (Johnsen et al. 1997), pollen records from Beringia indicate continued expansion of shrub tundra vegetation. Well-dated cores from central Alaska show that shrub tundra spread into upland areas during this interval (Bigelow and Edwards 2001; Bigelow and Powers 2001). By 13,000 calibrated years ago (13 cal ka), much of central Beringia had been inundated by rising sea levels (Manley 2002), which presumably increased available moisture over land areas (e.g., Mann et al. 2001:127–131). By 13.5 cal ka, mammoth was extinct in Beringia, except in isolated island refugia (Guthrie 2004, 2006).

The quantity and size of archaeological sites increases relative to the preceding interval, and their geographic distribution expands. Large occupations with complex features, including former dwellings and burials, appeared in southwestern Beringia on the Kamchatka Peninsula. Sites also appeared in northeastern Beringia, and upland areas were occupied probably for the first time.

The artifacts in these sites are difficult to interpret, however, and the final centuries of the Lateglacial interstadial have emerged as the most confusing but interesting period in Beringian prehistory. The wedge-shaped microblade cores and microblades that are present before 13.5 cal ka—and become common after 12.8 cal ka—are comparatively rare during this period. Assemblages containing various forms of bifacial points are more common. On Kamchatka, the points are typically stemmed and occasionally leaf-shaped, while in central Alaska, many of them are triangular or teardrop-shaped. Some generic lanceolate forms may be present in northwestern Alaska. Uncertainties about the classification of these assemblages have fueled debate among archaeologists over their place in Beringian prehistory and relationship to sites outside Beringia (Dumond 2001; Holmes 2001).

Paradoxically, the expansion in archaeological sites is accompanied by a decline in the quantity of preserved faunal remains relative to the preceding period. The available data nevertheless indicate a continuation of the economic pattern observed in the earlier Lateglacial sites. The diet remained focused on postglacial large mammals such as elk and sheep, with a heavy emphasis on small mammals and birds, especially waterfowl (Cook 1969; Goebel and Slobodin 1999). Sites continue to reveal traces of bone fuel (with woody shrubs), which probably consumed much of the food debris on occupation floors.

The artifact assemblages may reflect the ecological, rather than geographic, isolation of Beringia during this interval. According to the sea level and glacial chronology, a land connection between Chukotka and Alaska was still present and an ice-free corridor offered access to the northern Plains (Manley 2002; Dyke 2004). People now were living in both the adjoining areas of Northeast Asia and the northern Plains, but their artifacts and way of life were different from those of the Beringians, who apparently had created a unique Beringian tradition.

Kamchatka: Ushki

The most important sites in western Beringia lie on the Kamchatka Peninsula (figure 5.1). As briefly mentioned in chapter 4, the Ushki sites contain a stratified record of settlement in southwest Beringia from 13 cal ka to the inundation of the Bering Strait and beyond (Goebel,

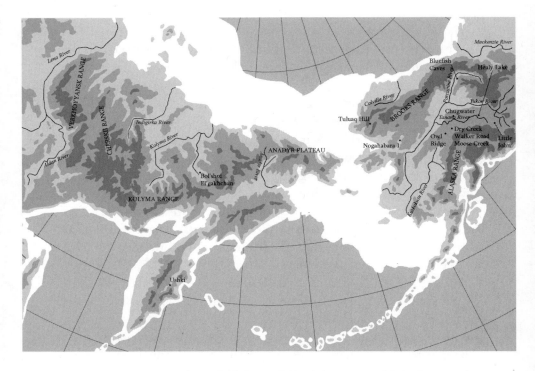

Figure 5.1 Beringia at the end of the Lateglacial (13.5–12.8 cal ka), showing the location of sites mentioned in the text

Waters, and Dikova 2003). Occupation horizons dating between 13 cal ka and 12 cal ka contain the most complex features known among any Beringian sites, including traces of multiple dwelling structures, pits, and a burial (Dikov 1977:47–58). Moreover, the lowest horizon at Ushki provided evidence of a stone tool industry that produced stemmed bifacial points—but apparently no microblades—and stimulated speculation about cultural connections between Beringia and northern North America (e.g., Dikov 1979:47–53).

The Ushki sites were discovered by the late N. N. Dikov in 1961–1962, and he worked at the sites each year between 1964 and 1990 (Dikov 1996:244). Despite the depth of the lowest archaeological levels (more than 2 meters at Ushki V), Dikov exposed hundreds of square meters of occupation area (Dikov 1977:53, fig. 11).[1] In 2000, Ted Goebel and several colleagues inaugurated a new research program at Ushki designed specifically to address questions about the geomorphic setting and age of the earliest components (Goebel, Waters, and Dikova 2003:502).

The five sites are found on the southern shore of Great Ushki Lake in central Kamchatka at latitude 55°N. The lake, which formed in late Holocene times, represents a cut-off meander of the Kamchatka River whose main channel is now situated north of the lake (Dikov 1977: 43–45). River terraces are found at elevations of 3 and 9 meters, respectively, above the modern channel margin. The archaeological sites rest on the second or higher terrace level and are buried in alluvial and eolian deposits (Goebel, Waters, and Dikova 2003:502–503). A sequence of Holocene volcanic ashes in these deposits provides support for dating and cross-correlation among the sites.[2]

The lowermost archaeological horizon (Layer VII) is found at Ushki I and Ushki V in an upward-fining sequence of alluvial silt and silty clay.[3] At the time of occupation, both sites rested on an aggrading floodplain and possibly the upper portion of a point bar (Goebel, Waters, and Dikova 2003:502). A new series of nine radiocarbon dates on charcoal and wood samples collected from four stratigraphic profiles (two at each locality) indicates that Layer VII dates between 13.2 and 12.7 cal ka (Goebel, Waters, and Dikova 2003:503, fig. 3). Because nearly identical dates were obtained on both the base-insoluble and soluble fractions of the charcoal samples, Goebel, Waters, and Dikova (2003:502) concluded that the latter had not been contaminated by older soluble organics (e.g., by groundwater containing old carbon) (figure 5.2; table 5.1).

Palynological analysis of samples collected from the alluvium indicates a relatively sparse vegetation cover dominated by tundra flora (Shilo, Dikov, and Lozhkin 1967:34–35). Substantial quantities of burned bone are reported from Layer VII (e.g., Dikov 1979:33; Goebel, Waters, and Dikova 2003:503), suggesting a continued reliance on bone fuel despite the warming trend and a comparatively mild local setting.

In 1974, the remains of a dwelling structure occupying more than 100 square meters were uncovered in Layer VII at Ushki I (Dikov 1977:46–50). The floor of the former dwelling extended 20–30 centimeters below the occupation surface and exhibited a figure-eight pattern that reflects a dual-chambered structure. Each chamber contained traces of several small pits and former hearths, along with stone implements, lithic flaking debris, grinding stones (sandstone), burned and unburned bone, and large quantities of red ocher and hematite (Dikov 1977:49, fig. 9). The artifacts also included three stone pendants.

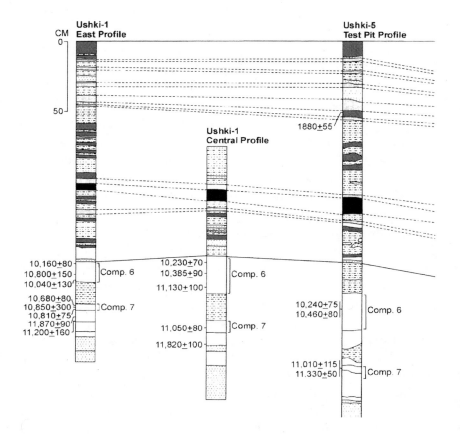

Figure 5.2 Stratigraphy of Ushki I and V (reprinted with permission from Goebel, Waters, and Dikova 2003:503, fig. 3, copyright 2003 AAAS)

With the exception of an antler of moose (*Alces alces*), none of the faunal remains was identified (Dikov 1977:50).

A second, slightly smaller but similar dual-chambered structure was encountered roughly 20 meters east of the first in 1979 (Goebel and Slobodin 1999:133, fig. 16). The contents also were similar to the other structure and included stone implements, waste flakes, grinding stone, and traces of hematite and ocher. Although none of the bones is identified to species, avian gastroliths (i.e., gizzard stones) are reported (Dikov 1990).[4]

Traces of at least four possible single-chambered dwellings also were mapped on Layer VII at Ushki I. One of these was excavated in the northeast quadrant of excavations in 1964 and occupied an extended oval area of roughly 60 square meters (Dikov 1977:50). A large, thick

Table 5.1

Calibrated Radiocarbon Dates from Sites in Beringia: Later Lateglacial
Interstadial (ca. 13.5–12.8 cal ka)

Site	Level	Material Dated[a]	Lab No.	Uncalibrated Date	Calibrated Age[b]
Ushki 1	Layer VII	Charcoal	MAG-637	9,750 ± 100	11,073 ± 168
	Layer VII	Charcoal	GIN-167	13,600 ± 250	16,608 ± 587
	Layer VII		MAG-522	13,800 ± 500	16,709 ± 858
	Layer VII		MAG-550	14,200 ± 700	17,163 ± 1015
	Layer VII	Charcoal	GIN-168	14,300 ± 200	17,670 ± 267
	Layer VII	Charcoal	AA-45710	10,675 ± 75	12,656 ± 77
	Layer VII	Charcoal	AA-45709	10,850 ± 320	12,669 ± 406
	Layer VII	Charcoal	AA-45708	10,810 ± 75	12,793 ± 80
	Layer VII	Charcoal	AA-45716	11,050 ± 75	12,946 ± 112
	Below Layer VII	Charcoal	AA-45711	11,870 ± 90	13,782 ± 159
	Below Layer VII	Charcoal	AA-45712	11,200 ± 160	13,097 ± 183
	Below Layer VII	Charcoal	AA-45718	11,820 ± 100	13,736 ± 159
Ushki 5	Layer VII	Charcoal	AA-41388	11,010 ± 115	12,932 ± 124
	Layer VII	Humates	AA-41389	11,050 ± 75	12,946 ± 112
	Layer VII	Charcoal	CAMS-74639	11,330 ± 50	13,222 ± 106
Healy Lake	Level 8	Bone	GX-1341	11,090 ± 170	13,014 ± 179
	Level 9	Charcoal	GX-2159	6,645 ± 280	7,515 ± 265
	Level 9	Charcoal	SI-738	8,210 ± 155	9,162 ± 207
	Level 9	Charcoal	AU-2	9,401 ± 528	10,777 ± 775
	Level 10	Charcoal	GX-2175	8,465 ± 360	9,482 ± 470
	Level 10	Charcoal	SI-739	10,040 ± 210	11,685 ± 361
	Level 10	Charcoal	AU-3	10,434 ± 279	12,159 ± 432
	Level 10	Charcoal	GX-1944	10,500 ± 280	12,233 ± 416
Dry Creek	Loess 2	Charcoal	SI-2880	11,120 ± 85	13,025 ± 140
Walker Road	Component I	Charcoal	AA-1683	11,010 ± 230	12,970 ± 200
	Component I	Charcoal	AA-1681	11,170 ± 180	13,070 ± 180
	Component I	Charcoal	AA-2264	11,300 ± 120	13,207 ± 148
	Component I	Charcoal	Beta-11254	11,820 ± 200	13,750 ± 240
Moose Creek	Component I	Charcoal	Beta-96627	11,190 ± 60	13,091 ± 117
Owl Ridge	Component I	Charcoal?	Beta-11079	2,380 ± 90	2499 ± 148
	Component I	Charcoal	Beta-5416	9,069 ± 410	10,255 ± 549
	Component I	Charcoal	Beta-11209	11,340 ± 150	13,253 ± 178
Tuluaq Hill	—	Charcoal	Beta-133394	7,950 ± 40	8,827 ± 113
	—	Charcoal	Beta-122323	11,110 ± 80	13,000 ± 90
	—	Charcoal	Beta-159913	11,120 ± 40	13,020 ± 50
	—	Charcoal	Beta-159915	11,160 ± 40	13,080 ± 80
	—	Charcoal	Beta-122322	11,180 ± 80	13,090 ± 100
	—	Charcoal	Beta-159914	11,200 ± 40	13,140 ± 70
	—	Charcoal	Beta-133393	11,200 ± 40	13,140 ± 70

SOURCES: Goebel and Slobodin 1999:109–110, table 1; Goebel, Waters, and Dikova 2003:503, fig. 3;
Cook 1996:327, table 6-8; Erlandson et al. 1991; Bigelow and Powers 2001:177–179, table 3; Pearson
1999:340, table 2; Phippen 1988:92, table 2; Rasic 2003:24, table 1
    [a] Blank space indicates data not available; dash indicates data not applicable
    [b] Dates calibrated with the Cologne Radiocarbon Calibration Programme CalPal-SFCP-2005
glacial calibration curve (September 2005) (www.calpal.de)

hearth, measuring 3 by 3 meters and containing burned bone fragments, was found on the northwest margin of the former structure. The dwelling contained stone tools, flakes, beads, pendants, and some ocher. Traces of another single-chamber structure, also containing a large hearth, were found nearby in 1978 (Dikov 1990). In addition to stone implements and ocher, this structure yielded burned bones of mammals and birds (Goebel and Slobodin 1999:134, table 2) (figure 5.3).

In 1989, traces of two more single-chambered structures were uncovered on the southern margin of the excavated area. Both were subcircular in outline with large central former hearths and contained stone implements and flaking debris (Dikov 1996:247; Goebel and Slobodin 1999:134, table 2).

The features in Layer VII excavated by Dikov also included a human burial pit, which remains unique among the archaeological sites of Beringia.[5] Discovered in 1964, the burial is located on the north-central area of the excavations and not in close proximity to any of the former dwelling structures. The pit is circular in plan with a diameter of 1.8 meters and a depth of 70 centimeters (Dikov 1969:95–96). Although the human remains were poorly preserved—the bones apparently observable only in outline form—numerous artifacts were found in the pit. The latter included 881 beads of pyrophyllite and 26 chipped points of chalcedony. Dikov (1969:96) suggested that one group of 20 beads arranged in a diamond shape might represent ornaments sewn into the clothing of the deceased (Dikov 1967:25). The pit contained a number of large rocks, rounded cobbles, and substantial quantities of red ocher. Red ocher was also found in sediment surrounding the pit (Dikov 1969:95).

Layer VII is found at Ushki V as well, although Dikov (1977:80–82) reported only isolated artifacts from this locality (e.g., stemmed point). Goebel, Waters, and Dikova (2003:503) focused their excavations in 2000 on Ushki V and exposed an area of 20 square meters that contained two former hearths. Both hearths were approximately 1 meter in diameter and yielded charcoal and burned bone fragments.

Despite the impressive features, it is the stone artifact inventory from Layer VII that has received the most attention from archaeologists. The assemblage lacks wedge-shaped microblade cores, microblade-core parts, and—apparently—microblades (Dikov 1967:23, fig. 6).[6] It also lacks multifaceted burins, which are closely associated with microblade

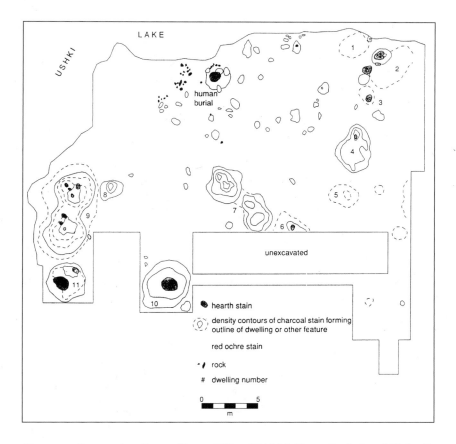

Within the figure:

LAKE

USHKI

human burial

unexcavated

1

2

3

4

5

6

7

8

9

10

11

● hearth stain

(○) density contours of charcoal stain forming outline of dwelling or other feature

red ochre stain

• ▮ rock

dwelling number

0   5

m

Figure 5.3 Occupation floor of Layer VII at Ushki I (from Goebel and Slobodin 1999:133, fig. 16; reprinted with permission from the Center for the Study of the First Americans. Copyright 1999)

technology in Siberia and Beringia. The assemblage contains bifacial stemmed points that, although not unique in Beringia, are firmly dated to the 13 cal ka time range only at Ushki. Dikov (1996:247) recovered a total of fifty stemmed points during his excavations, while Goebel, Waters, and Dikova (2003:503–504) retrieved another five points from their 2000 excavation at Ushki V. The points were manufactured on flakes and not always fully flaked on both sides (Dikov 1979:34–35; Goebel, Waters, and Dikova 2003:503) (figure 5.4).[7]

Other elements of the chipped stone assemblage include several leaf-shaped or teardrop-shaped bifacial points, as well as larger bifaces. Although multifaceted burins are absent, other types of burins, including

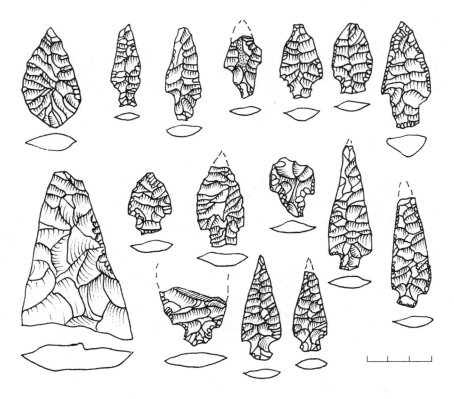

Figure 5.4 Artifacts recovered from Layer VII at Ushki (after Dikov 1979:35, fig. 3)

diagonal burins on blade-like flakes and burins on bifacial and unifacial tools, are present. Also found in Layer VII were end-scrapers and unifacial "knives" (Dikov 1979:34–38). A series of small perforators or awls were recovered from the burial pit (Dikov 1967:29, fig. 12).

From early publications onward (e.g., Dikov 1967:30–31), Dikov maintained that Layer VII contained a set of remains that could be differentiated from those of the overlying layers at Ushki. This conclusion was based primarily on the sharply contrasting stone artifact assemblages recovered from Layer VII and overlying layers (see chapter 6), but other differences—such as the shape of the dwellings, design of the hearths, and contents of the burial—also are apparent (Dikov 1996: 245–247). He disputed the suggestion of Mochanov that Layer VII could be considered part of the Dyuktai culture (Dikov and Titov 1984:74). Sites in other parts of Beringia—especially eastern Beringia— contain remains either firmly or tentatively dated to the same time period that also suggest the presence of a different industry, although close parallels with Ushki Layer VII are rare.

The Layer VII occupation at Ushki almost certainly reflects a relatively long-term encampment. This is indicated by the traces of dwelling structures, thick hearth layers of burned bone and charcoal,[8] diverse stone tool assemblage, and large quantities of lithic debris (Dikov 1979:33). In fact, there is nothing comparable in terms of size and complexity known thus far in the Beringian archaeological record—with the exception of the overlying horizon (Layer VI) at Ushki I (Dikov 1977: 52–58). The Kamchatka region, which today enjoys a comparatively mild and wet climate as well as high biotic productivity, may have consistently supported larger human groups than other parts of Beringia.

Although few identifiable faunal remains were reported, Dikov (1967:30) speculated that the Layer VII settlement was tied to a broad-based economy. The presence of bird bones and avian gastroliths seems to support this view, and excavations at Ushki V in 2006 reported both bird bones (duck) and fish remains (salmon) from Layer VII (Ponkratova 2006). (The overlying occupation horizon [Layer VI] also yielded bones of birds and fish [Dikov 1996:245; Goebel, Waters, and Dikova 2003:503]). The pattern appears similar to that of the earlier sites (described in chapter 4) and further underscores the lack of evidence for specialized hunting of late Pleistocene large mammals. In contrast, the reports of burned bone from Layer VII (e.g., Dikov 1979:33; Goebel, Waters, and Dikova 2003:503) suggest once again that many large mammal remains may have been consumed as fuel.

Chukotka

At least one archaeological site in Chukotka contains remains that may be contemporaneous with the lowest occupation horizon at Ushki, but the dating is based primarily on typologic comparisons and is therefore problematic. Bol'shoi El'gakhchan I is found in an upland setting more than 370 meters above sea level, along the Omolon River on the northern slope of the Kolyma Range (Kir'yak 1993:16–17; T. Goebel, pers. comm., 2005). If the lower occupation level actually dates to ca. 13 cal ka, Bol'shoi El'gakhchan would have some parallel to the sites of comparable age in the northern foothills of the Alaska Range (described below).

The site is located on a bedrock promontory at the confluence of the Bolshoi El'gakhchan River and the Omolon. The bedrock, which is more than 36 meters above the modern floodplain, may be mantled with alluvium deposited during the Last Glacial (Kir'yak 1996:228–229). Artifacts assigned to the Neolithic are buried in the modern soil layer (up to 14 centimeters below the surface) that caps a deposit of sandy loams overlying the alluvium. An older assemblage is buried roughly 40 centimeters beneath the surface in the sandy loam, which exhibits traces of frost disturbance (some artifacts were found in vertical position) (Kir'yak 1993:18–19). A pollen sample from the sediments containing the older assemblage reportedly indicates pre-Holocene vegetation (Kir'yak 1996:235), but no radiocarbon dates are available for the site.

The lower occupation level produced four stemmed bifacial points similar to those found in Layer VII at Ushki. A fifth point—leaf-shaped— is also similar to one from the lowest level at Ushki, as well as to bifacial points recovered from contemporaneous sites in central Alaska (Kir'yak 1993:19–23; Goebel and Slobodin 1999:124). The most common items in the assemblage are retouched and unretouched large blades, while the second most numerous stone artifacts are side-scrapers and end-scrapers (Kir'yak 1996:230–233). Burins are rare and represented by atypical forms. Some microblades are present, but they were found on the margin of the occupation area and in shallow sediment near the bluff edge, and are thought to be unrelated to the lower component (Kir'yak 1996:230).

Because of the similarities with Ushki Layer VII and several central Alaskan sites dating to ca. 13 cal ka (e.g., Walker Road), the older assemblage at Bol'shoi El'gakhchan I might represent an occupation in the same time range (Kir'yak 1996:235). In the absence of supporting dates, however, this remains tentative (Goebel and Slobodin 1999:124).

Tanana River Valley

As noted in chapter 4, the three sites near the mouth of Shaw Creek in the Tanana Valley constitute the most important evidence of settlement in Beringia before 13.5 cal ka. These sites—Broken Mammoth, Swan Point, and Mead—also contain occupation levels that date to the Younger Dryas period (described in chapter 6). Curiously, they lack definitive traces of occupation during the final centuries of the

Lateglacial (i.e., 13.5–12.8 cal ka).[9] Two other sites in the Tanana River Valley contain artifacts that probably date to this period, although only one of these assemblages has been dated directly by radiocarbon.

HEALY LAKE

Located near the mouth of the Gerstle River in the Upper Tanana Valley, the archaeological remains at Healy Lake were discovered in 1962 by Robert A. McKennan in the course of ethnographic research in the region. Testing in 1966 revealed buried occupation layers in several locations, and excavations were undertaken by John P. Cook during 1967–1972. Most evidence of early occupation was found at the Village site, where approximately 400 square meters were eventually exposed (Cook 1969:7–9, 1996:323). The results of these excavations—partly described in Cook's 1969 doctoral thesis—greatly influenced Beringian archaeology in the decades that followed.

Draining the Yukon-Tanana Upland, the Tanana River has dammed a series of small lakes on the north side of the valley upstream of Fairbanks. At 345 meters above sea level, Healy Lake is one of the larger of these lakes and was the location of a historic Athapaskan village (Cook 1989). The area is rich in waterfowl and fish during the warmer months (Cook 1969:16–17). The Village site is found on a bedrock ridge that projects into the lake from the eastern shore. Above the bedrock and an overlying layer of wind-abraded cobbles and gravels lies a sterile layer of coarse sand containing a band of pebbles (Cook 1969:59–60). The sand is capped with a bed of upward-fining loess that varies between 0.6 and 1.2 meters in thickness and is, in turn, capped with the humic layer of the modern soil. Traces of buried soils are evident in the loess, and these are associated with the lower occupation levels (Cook 1996:323–325).

The buried artifacts and features lie beneath the modern humus, extending downward into the lower portion of the loess. During excavation, Cook subdivided them into ten arbitrary 5-centimeter levels, which were subsequently grouped into three "cultural stages" (Cook 1996:324–325).[10] The earliest of these stages—eventually named "Chindadn"—comprises the lowermost five levels (levels 6–10). Twenty-two radiocarbon dates have been obtained on burned bone, charcoal, and other materials from these levels, and they provide a wide range of ages between 13.4 and 5.9 cal ka (Erlandson et al. 1991; Cook 1996:327).

Thus, only a portion of the Chindadn horizon may date to the end of the Lateglacial and before the beginning of the Younger Dryas period.

Although Cook did not find traces of former dwellings, he encountered numerous hearths, as well as concentrations of rocks, flakes, and bone, in the lowest levels of the Healy Lake Village site (Cook 1969: 240–244). Large quantities of burned bone were recovered from the hearths (up to 10 cm in thickness). Noting that the amount of burned bone in the hearths decreased in the upper levels, Cook (1969:242–243) speculated that wood fuel was scarce during the Chindadn occupation. Although most bones in the hearths were not identifiable to species, they were attributable primarily to birds ("presumably waterfowl") and—to a lesser extent—to small and large mammals (hare, squirrel, and caribou or sheep) (Cook 1969:240–243, 1996:324). Both the lack of winter houses and the predominance of bird remains suggest a spring occupation.

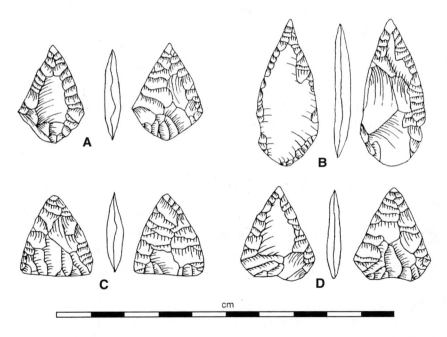

Figure 5.5 Bifacial points (Chindadn points) recovered from the lower levels of Healy Lake (from Cook 1996:326, fig. 6-11; reprinted from *American Beginnings: The Prehistory and Paleoecology of Beringia* with permission of Frederick Hadleigh West. Copyright 1996)

Stone artifacts recovered from the five levels assigned to the Chindadn horizon include two wedge-shaped microblade cores and ninety-two microblades, as well as small and thin bifacial points (subsequently termed "Chindadn points"). The points are either teardrop-shaped or triangular (figure 5.5). Also present in these levels are larger bifaces ("knives"), end-scrapers, and several other types, including a spoke-shave and burin (Cook 1969:66–117, 1996:325–326). No bone, antler, or ivory artifacts were found.

Because of the thickness of the Chindadn horizon and the wide range of dates—indicating a time span of several millennia—some authors have suggested that more than one type of assemblage might be combined in this unit (e.g., Hoffecker, Powers, and Goebel 1993:49; Hamilton and Goebel 1999:169).[11] Given the variety of industries that now appear to have been present in eastern Beringia during the Lateglacial and Younger Dryas (summarized in chapter 7), this remains a possibility.[12]

CHUGWATER

In 1975, John Cook discovered another early prehistoric site in the Tanana River Valley on a bedrock bluff roughly 35 kilometers southeast of Fairbanks (Lively 1988:1–3).[13] Moose Creek Bluff lies 67 meters above the present Tanana River floodplain and 224 meters above sea level. Much of the shallow sediment overlying the bedrock appeared to have been disturbed by earlier construction activity. Nevertheless, following additional testing at the Chugwater site,[14] large-scale excavations—ultimately exposing about 400 square meters—were performed during 1982–1987 (Maitland 1986; Lively 1988).

In places, the bedrock bluff is capped with a sand unit that is thought to date to the later phases of the Last Glacial Maximum (Thorson 1983). A layer of colluvium composed of rubble, sand, and silt overlies the sand and the bedrock, and this is capped with a unit of loess that generally varies between 25 and 50 centimeters in thickness and which underlies the modern soil (Lively 1996:309, fig. 6-3). As at Healy Lake, traces of a buried soil containing multiple weakly developed A horizons are visible in the loess and probably represent an early Holocene or terminal Pleistocene soil. A total of twenty radiocarbon dates from the loess suggest that the latter spans the entire Holocene—and probably the terminal Pleistocene—although most of the dates are younger than

2,600 radiocarbon years and overall exhibit a poor correlation with depth (Hamilton and Goebel 1999:169).

Three cultural horizons were defined at Chugwater, but only the upper two horizons yielded radiocarbon dates. Two dates tentatively associated with the middle horizon suggest an age of ca. 10.8–10 cal ka (Lively 1996:310). The lower horizon thus may antedate 11 cal ka, and comparison of the artifacts with Healy Lake and dated assemblages from the North Alaska Range suggest that it might be as old as 13 cal ka (Lively 1996:310–311; Hamilton and Goebel 1999:169). The artifacts include small bifacial points (i.e., Chindadn points), end-scrapers, and a bifacial knife fragment. However, the shallow sediments, severe disturbance, and problematic association between the dates and the cultural horizons preclude firm conclusions about the early occupation at Chugwater.[15]

North Alaska Range

In 1973, Charles E. Holmes discovered an archaeological site near the town of Healy, Alaska, during a survey of the newly completed Parks Highway corridor (Holmes 1974). The site was located on a bluff overlooking Dry Creek in the Nenana Valley, and its subsequent investigation stimulated much additional research in the northern foothills of the Alaska Range and had a significant effect on Beringian archaeology (Powers and Hoffecker 1989). At Dry Creek and other localities in the northern foothills, archaeological remains were found in deep eolian deposits that provided a superior stratigraphic context (Thorson and Hamilton 1977; Hoffecker 1988).

The windy upland valleys of the foothills zone were originally viewed as places that would have preserved remnants of the herbaceous steppe flora and large grazing mammals of the LGM—and thus especially attractive to hunters of the Lateglacial period (Ager 1975:85–86; Guthrie 1983a:215–216; Powers and Hoffecker 1989:283–284). As elsewhere in Beringia, however, occupation of these valleys seems to coincide with the spread of shrub tundra vegetation into the foothills (Bigelow and Powers 2001:182–183), and it may have been the presence of woody plants like willow and dwarf birch that brought people to higher elevations at this time. The high-resolution pollen and lake level record from Windmill Lake—located in the upper Nenana Valley at 615 meters above sea level—indicates a rapid rise in birch pollen and lake level at about 13.5 cal ka (Bigelow and Edwards 2001:208–211).

Three sites in the Nenana valley were eventually found to contain occupation levels dating to the end of the Lateglacial period. In addition to Dry Creek, Walker Road—located on the opposite side of the valley—produced remains from this period. And downstream on a high terrace, Moose Creek yielded an occupation level dating to ca. 13.1 cal ka (Powers and Hoffecker 1989; Pearson 1999). At least one site in the neighboring Teklanika Valley, Owl Ridge, also contains remains of comparable age (Hoffecker, Powers, and Phippen 1996).

The North Alaska Range sites reveal a new aspect of the Beringian Lateglacial economy. As in later times, people probably moved into the upper valleys during the autumn when various game species were concentrated in the foothills (Guthrie 1983a:285). Small mammals and birds may have been less heavily exploited at these sites than they were in the Tanana Basin and other lowland areas. Nevertheless, while the remains of sheep and wapiti were identified in the lowest occupation level at Dry Creek (13 cal ka), avian gastroliths were also recovered (Guthrie 1983a:274–281).

DRY CREEK

The Dry Creek site is found on an ancient glaciofluvial outwash terrace at the upper end of the foothills segment of the Nenana Valley, roughly 470 meters above sea level. The outwash was deposited during a pre-LGM glaciation (locally termed "Healy") and subsequently incised by the river—and by Dry Creek—forming a terrace and bluff (Wahrhaftig 1958; Thorson and Hamilton 1977:167–172). The south-facing location of the site affords a view of the upper valley and the mountain front (figure 5.6).

At some point after 15–14 cal ka, eolian sand and silt began to accumulate on the gravel surface of the Healy outwash terrace.[16] Above a basal unit of sterile loess lies another loess layer that contains the earliest occupation level—dated to 13.1 cal ka (Thorson and Hamilton 1977:156, table 1). The latter is capped with a thin and discontinuous sand layer, which, in turn, is overlain with another artifact-bearing loess unit with an age range of 12.5–11.6 cal ka. The younger loess unit also contains

Figure 5.6 Topographic setting of the Dry Creek site (photograph by W. R. Powers)

a series of thin organic lenses that may represent incipient tundra soil formation. More layers of loess and buried soils dating to the early Holocene overlie this unit. The upper part of the sequence comprises eolian sandy silts and sands of the later Holocene containing buried forest soils (Thorson and Hamilton 1977:161–162) (figure 5.7).

Most of the archaeological research at Dry Creek was undertaken by the late William Roger Powers, who directed excavations there during 1974 and 1976–1977 (Powers, Guthrie, and Hoffecker 1983). More than 345 square meters were exposed, and features—traces of hearths and concentrations of occupation debris—were mapped across floor areas for both lower components (Powers and Hoffecker 1989:277–282). Three large (2–5 square meters) and several smaller concentrations were found in the lowest level, including a small cluster of decomposed faunal remains. As noted above, identifiable mammal remains (teeth only) included Dall sheep (*Ovis dalli*) and wapiti (*Cervus canadensis*), while the avian gastroliths fall in the ptarmigan size range (mean of 2.14 millimeters) (Guthrie 1983a:274–281).

Former hearths in both components were represented by isolated fragments of charcoal (presumably from woody shrubs). An analysis of

Figure 5.7 Stratigraphic profile of the Dry Creek site, comprising a sequence of eolian silts and sands with several buried soils (photograph by W. R. Powers)

hearth areas revealed large quantities of opaline phytoliths from festucoid grasses (Guthrie 1983a:282–283). It seems unlikely that the occupants of Dry Creek were burning grass, and the phyloliths probably are derived from dung that was used for fuel. Alternatively, they may simply reflect abundant growth in these areas after the site was abandoned.[17]

More than 3,500 stone artifacts were recovered from the lowermost component. The assemblage contained a small bifacial point and two small bifacial point bases and a tip, end-scrapers, larger bifaces ("knives"), side-scrapers, and a large unifacial plane (Powers and Hoffecker 1989:281; Hoffecker, Powers, and Bigelow 1996:347–349). No microblade cores, core parts, or microblades were found in this occupation level (figure 5.8).

The absence of microblades—combined with the small bifacial point and point fragments—in the lowest level at Dry Creek encouraged the definition of a new complex (Nenana complex), although this was not done until a similar pattern had emerged from other stratified sites in the Nenana Valley (Powers and Hoffecker 1989:278). Some researchers concluded that the earliest component at Dry Creek represented a subsample of artifacts derived from the overlying component (e.g., Dumond 2001:199), although this did not account for the later discoveries in the valley. Others suggested that the lower level at Dry Creek and other Nenana complex assemblages were merely functional subsets of a larger industry or tradition that contained microblade technology (West 1996a:546–552).

In any case, Dry Creek seems to have been used as a short-term hunting camp rather than for protracted occupation as at Ushki (Powers, Guthrie, and Hoffecker 1983). It would seem to have functioned as both an overlook and a temporary camp. Stone tools were manufactured and resharpened at the site (Smith 1985), and perhaps some game was processed there as well. The high proportion of end-scrapers in the lowest occupation level may reflect a special emphasis on hide-working, which probably was an important aspect of the large mammal hunting at this site and others in the North Alaska Range.

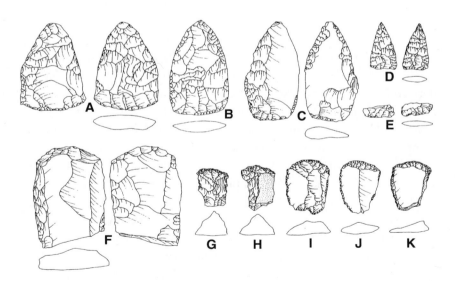

Figure 5.8 Artifacts from the lowest level (Component I) at Dry Creek, classified as the "Nenana complex" (from Powers and Hoffecker 1989:278–281; reproduced by permission of the Society of American Archaeology from *American Antiquity*, vol. 52, no. 2, 1989)

WALKER ROAD

The Walker Road site was discovered in 1980 but not investigated substantively until 1984 and then was excavated during 1985–1990 (Goebel et al. 1996:356). The site is located 10 kilometers north of Dry Creek (i.e., downstream) adjacent to a small and unnamed side-valley creek, at an elevation of 430 meters above sea level. As in the case of Dry Creek, artifacts and features are buried in loess deposits—roughly 1 meter in depth—that overlie a glaciofluvial outwash terrace of Healy age (Wahrhaftig 1970).

The stratigraphy of the eolian deposits at Walker Road is similar— although not identical—to that of Dry Creek (Goebel et al. 1996: 356–358). A basal unit of loess dates to slightly before 13 cal ka and contains the major archaeological layer at the site. Unlike the sites described above, Walker Road yields little evidence of middle and late Holocene occupation. The basal loess is capped with a sand layer that appears to correlate with the sand overlying the earliest archaeological level at Dry Creek. Above the sand lie two layers of loess. The upper loess contains a soil complex that yielded an AMS date on wood charcoal of 9.8 cal ka (Goebel et al. 1996:358) and may be correlated with a similar buried soil complex at Dry Creek (Hoffecker 2001:143). The sediments above the loess contain a buried forest soil of late Holocene age.

Numerous small fragments of burned bone are reported from the main occupation layer, although none was identifiable to taxon (Goebel et al. 1996:358). Several concentrations of debris were delineated across an exposed area of approximately 200 square meters . The largest of these comprised a bowl-shaped hearth (filled with both wood charcoal and burned bone) surrounded by a roughly circular distribution— with a 2-meter radius—of roughly 2,000 pieces of lithic debris and more than 150 tools (Goebel et al. 1996:359). On the north side of the hearth, the tools are chiefly confined to end-scrapers and flakes apparently struck from the edges of end-scrapers. On the opposite side of the hearth, a more diverse set of tools—side-scrapers, wedges, end-scrapers, and others—was found. The entire hearth and debris complex might represent the location of a light circular structure (Goebel and Powers 1989:7–9) (figure 5.9).

A smaller concentration was found near the terrace edge and contained another circular hearth of charcoal and bone associated with two anvil stones and roughly 1,000 flakes and 50 retouched items

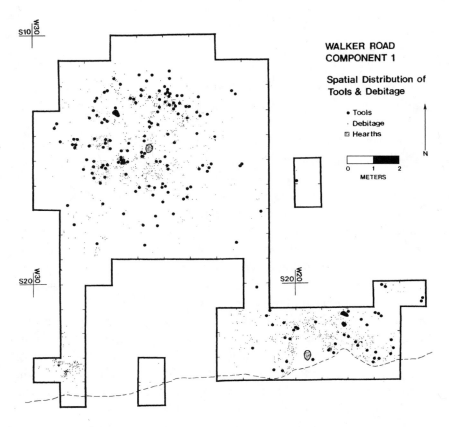

Figure 5.9 Occupation floor at the Walker Road site in the Nenana Valley (from Goebel and Powers 1989: fig. 11; reproduced with permission from Ted Goebel)

(Goebel et al. 1996:359). Among the tools were end-scrapers, side-scrapers, gravers, and other forms. Two other smaller concentrations lacking associated hearths are also present.

The artifact assemblage recovered from the main occupation layer at Walker Road numbered almost 5,000 items, including 218 retouched pieces (4.3 percent of the total). The assemblage is devoid of wedge-shaped microblade cores, microblade-core parts, and microblades, as well as burins (Goebel et al. 1996:360–362). Among the tools are end-scrapers on blades and flakes, side-scrapers, large scraper planes, wedges (or *pièces ésquillées*), gravers, and bifacial implements. The last category includes four small teardrop-shaped points (i.e., Chindadn points) and three larger specimens (figure 5.10).

The absence of microblade technology (and burins) is significant, because unlike Dry Creek or several other sites in this time range, the Walker Road assemblage was clearly not derived from an overlying occupation containing such technology. Walker Road is probably the most suitable "type site" for the Nenana complex (Powers and Hoffecker 1989). As at Dry Creek, the high proportion of end-scrapers may reflect an emphasis on the processing of large mammal hides.

MOOSE CREEK

Situated on the highest terrace surface in the Nenana Valley—more than 200 meters above the modern river floodplain and roughly 500 meters above sea level—Moose Creek almost certainly functioned as a hunting overlook that afforded a view of much of the central valley. The site was discovered in 1978 during efforts to find more sites in the North Alaska Range after the excavations at Dry Creek. Preliminary testing was undertaken in 1979 and 1984 (Hoffecker 1996:363), and more extensive investigation at Moose Creek was performed in 1996 (Pearson 1999).

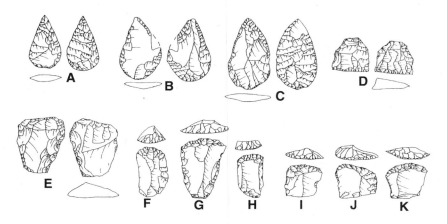

Figure 5.10 Artifacts recovered from the lower level at Walker Road and assigned to the "Nenana complex" (from Goebel et al. 1996:361, fig. 7-14; reprinted from *American Beginnings: The Prehistory and Paleoecology of Beringia* with permission of Frederick Hadleigh West. Copyright 1996)

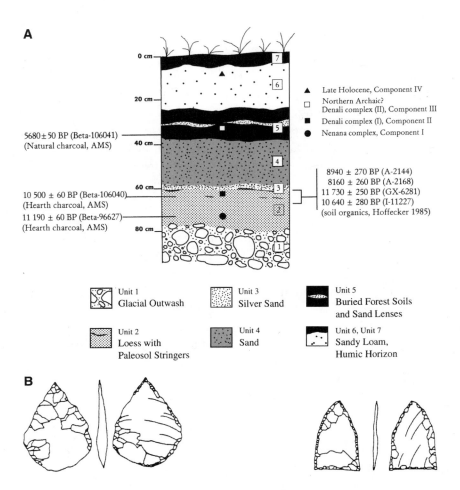

Figure 5.11 Moose Creek: (A) stratigraphy and (B) bifacial points from the lowest level (from Pearson 1999:334, 337, figs. 26e and 26f; reproduced from *Arctic*, vol. 52, no. 4, with permission of the Arctic Institute of North America. Copyright 1999)

The highest terrace level represents the surface of the uplifted and poorly consolidated Tertiary rock that constitutes the north-central foothills of the Alaska Range (Wahrhaftig 1958:15–16). At Moose Creek, this Tertiary rock is capped with a rare deposit of ancient glacial till. The archaeological remains are buried in eolian sand and silt—generally less than 1 meter in depth—that unconformably overlies the till. The lowest occupation level at Moose Creek is contained in a basal unit of loess that dates to 13.1 cal ka. A younger occupation level has been identified in the upper portion of the loess unit; it is associated with traces of tundra soil development and dated to 12.4 cal ka. Sand and sandy loam units are deposited above the loess, and they contain two later components associated with forest soil formation (Pearson 1999:336).

A total of approximately 65 square meters was excavated at Moose Creek—including several large test pits beyond the margins of the main excavation units—and it revealed a comparatively small living area (Pearson 1999). No large concentrations of debris were found in the lowest occupation level, but traces of a hearth produced some wood charcoal identified as willow (*Salix* spp.) (Pearson 1999:339). Retouched stone artifacts from this level—classified as part of the Nenana complex—were confined to two small bifacial points (Chindadn points), one larger biface, one side-scraper, one scraper-plane, and a retouched flake. The Chindadn points include teardrop-shaped and subtriangular forms, both of which exhibit incomplete bifacial flaking (Pearson 1999) (figure 5.11).[18]

OWL RIDGE

At least one other valley in the northern foothills of the Alaska Range may contain traces of occupation dating to about 13 cal ka. The Teklanika Valley, which lies 10–15 kilometers west of the Nenana River, yielded a small site on a south-facing glacial outwash terrace at roughly the same elevation as Dry Creek (470 meters above sea leavel). Owl Ridge was primarily investigated by Peter Phippen in 1982 and 1984, who excavated 26 square meters at the site (Phippen 1988; Hoffecker, Powers, and Phippen 1996:353).

Three occupation levels were identified in an eolian sequence similar to that of the Nenana Valley sites. The lowest level is buried in a loess unit that unconformably overlies the glacial outwash and is capped with a sand horizon of varying thickness. A radiocarbon date from this level suggested an age of 13.3 cal ka, but two younger dates were also reported from the same layer (Phippen 1988:92, table 2). The middle occupation level, which is associated with a buried soil complex similar to the lower soils at Walker Road and Dry Creek, yielded four dates of ca. 10.5–8 cal ka. Fewer than 150 artifacts were recovered from the lowest component, including three bifaces and two biface fragments. The latter seem to represent lanceolate point bases (Phippen 1988).

Although the lowest component at Owl Ridge has been compared with the earliest assemblages in the Nenana Valley and was tentatively added to the Nenana complex (e.g., Phippen 1988:137; Powers and Hoffecker 1989:284), the artifacts are of ambiguous affiliation (Hoffecker, Powers, and Phippen 1996:355). Combined with uncertainties about the dating of the basal loess, the status of the component is problematic. It may actually belong to a group of assemblages containing lanceolate points that are dated in the North Alaska Range to the Younger Dryas interval (see chapter 6).

Northern Alaska

In 1975, a radiocarbon date of 13.5 cal ka was reported from a former hearth at the site of Putu in the northern Brooks Range (Alexander 1987:42). The date generated intense interest among archaeologists because it was associated with a fluted point. It also provided the first evidence of Lateglacial period settlement in northern Alaska. Subsequent analysis of laboratory records revealed that this date actually was obtained on a sample from sediments below the occupation level (Reanier 1996:509).

In recent years, evidence of Lateglacial occupation in northwestern Alaska has emerged from Tuluaq Hill, located on a small tributary of the Kelly River, which drains the southern slope of the De Long Mountains in the western Brooks Range. A large assemblage of stone artifacts was recovered from the surface and a shallow stratigraphic context, along with several former hearths. The hearths contained flaking debris and pieces of willow or poplar charcoal that yielded five radiocarbon dates of 13.1–13 cal ka and one younger date of 8.8 cal ka (Rasic and Gal 2000:66–67; Rasic 2003:24).

The artifact assemblage from Tuluaq Hill comprises 64 projectile points and point fragments and more than 300 biface fragments. The points are lanceolate in form with rounded bases, and many exhibit impact fractures. Other artifacts include end-scrapers, thick unifacial tools, gravers, and some retouched flakes. Among the artifacts recovered from the surface were a fluted point and wedge-shaped microblade core (Rasic and Gal 2000:66–67; Rasic 2003:24). The artifacts were produced on a local chert that crops out within a few kilometers of the site.

Because of the shallow stratigraphy, the assemblage may represent a mixture of artifacts from different time periods (Rasic 2003:24). Similar lanceolate points ("Sluiceway Points") have been dated to a somewhat younger range (ca. 11.6–10.9 cal ka) in the Noatak River area (Rasic 2003), and both fluted points and gravers are found in north Alaska sites of the Younger Dryas interval (see chapter 6). Nevertheless, the early dates from the hearths indicate a human presence, and the points from Tuluaq Hill may be tentatively included among the variety of Beringian point types that have been found in sites of the later Lateglacial interstadial.

New evidence of occupation in this time range has been reported from Nogahabara 1, located in the Nogahabara Sand Dunes area in the Koyukuk River Basin. Information about the site is currently limited, but it is reportedly buried in eolian sand (dune deposit) and contains a former hearth that yielded a date of ca. 13 cal ka (Odess and Rasic 2004). The associated artifacts are primarily of obsidian from the nearby Batza Téna source and comprise a variety of types, including microblades and various bifacial point forms. The presence of side-notched points—which are dated elsewhere in Alaska to the early mid-Holocene—in the assemblage suggests that younger time periods also are represented in the single component at the site.

Yukon Territory

Two sites in the Yukon Territory may date to the end of the Lateglacial, but the dating of both remains uncertain. The first of these sites is Bluefish Caves in the northern Yukon, which is described in chapter 4. As noted, the three caves are found at 600 meters above sea level on the southern edge of the Bluefish Basin. A small assemblage of stone artifacts—associated with large mammal bones, as well as remains of smaller vertebrates—was recovered from a loess unit dating to the LGM and later (Morlan and Cinq-Mars 1982:368).

Most of the artifacts were concentrated in the zone where pollen samples indicated the shift to shrub tundra (Cinq-Mars 1990:21–22), following the pattern seen elsewhere across Lateglacial Beringia. Current dates from pollen sequences in Alaska suggest that the shift took place later than previously thought (Bigelow and Powers 2001:187–188). At elevations over 400 meters above sea level in south-central Alaska, shrub tundra did not replace herbaceous tundra until ca. 13 cal ka (Bigelow and Edwards 2001:211), and similar dates are reported from pollen sequences in the Mackenzie Mountains (Szeicz and MacDonald 2001:250–253). Accordingly, 13 cal ka seems likely to be a maximum date for most if not all of the artifacts at Bluefish Caves. If the assemblage does date to this period, it would represent an example of microblade technology in Beringia during the interval 13.5–12.8 cal ka.

A newly discovered site in southwestern Yukon—near the Alaska border—also may contain an occupation dating to the end of the Late-glacial. Little John (KdVo-6) is found north of Beaver Creek in the up-per Mirror Creek Valley. The area lies in the uppermost Tanana River drainage and was perhaps first visited by people traveling up the river from places like Healy Lake. Norman Easton conducted excavations here in 2002–2004 and 2006 and has reported preliminary results (Easton et al. 2004; N. A. Easton, pers. comm., 2006).

The Little John site rests on an ancient moraine (Mirror Creek glaciation) that correlates with the Healy/Delta glaciation in the North Alaska Range (Wahrhaftig 1970). In the eastern area of the site, eolian deposits overlying the till are roughly 1 meter in thick-ness. Below the modern soil and a volcanic tephra lies roughly 60 centimeters of pebbly loess. Underneath the loess are two buried soil horizons—separated by loess—that yielded a large quantity of well-preserved fauna, including bones of caribou, hare, swan, and small birds. Two radiocarbon dates were obtained on bone from the upper buried soil (10.9 and 9.9 cal ka). Two small bifaces and some flaking debris were recovered in association with the fauna, and they are assigned to the Denali complex (N. A. Easton, pers. comm., 2006).

In the western area of the Little John site, only 30 centimeters of loess overlie the till. Below a horizon of microblades (also assigned to Denali), Easton recovered an assemblage containing large bifaces and Chindadn points, which he assigned to the Nenana complex (Easton et al. 2004). The artifacts in the western potion of the site remain un-dated. The site is analogous in many respects to Chugwater in the lower Tanana Basin.

Summary and Conclusions

During the final centuries of the Lateglacial interstadial, human settlement expanded into new areas of Beringia. The number and size of sites dating to this interval may or may not indicate some increase in population but almost certainly reflect movement into higher elevations and new habitats. For the first time, locations higher than 300 meters above sea level were occupied at places like Healy Lake in the Upper Tanana, Dry Creek and Walker Road in the North Alaska Range foothills, and probably Bluefish Caves in the northern Yukon.

At most if not all of these upland locations, occupation coincides with the arrival of shrub tundra vegetation in the area. As in the case of the earlier settlement of lowland basins in Beringia, the presence of woody shrubs such as willow and dwarf birch may have been the critical variable—as starter fuel for bone (table 5.2). Full-sized trees remained scarce in Beringia, and burned bone is reported from most of these sites. But movement into upland areas may have presented people with a new problem: a scarcity of large natural concentrations of bone, which seem to have accumulated primarily in lowland basins such as the Kolyma and Tanana (Péwé 1975:98; Vereshchagin and Baryshnikov 1984). At sites such as Dry Creek and Walker Road, bone fuel sources may have been restricted chiefly to hunted prey remains, and their

Table 5.2
Plant Taxa Identified in Beringian Archaeological Sites

Site	Level	Identified Taxa	Calibrated Age (cal ka)
Berelekh	Associated with artifacts	*Salix* spp., *Larix* spp.	14
Swan Point	Cultural Zone 4	*Salix* spp., *Salix/Populus*	14
Moose Creek	Component I	*Salix* spp.	13.1
Tuluaq Hill	Associated with lithics	*Salix* spp./ *Populus*	13.1
Dry Creek	Component II	Festucoid grasses	12.5
Mesa	—[a]	*Salix* spp., *Populus, Alnus*	11.5
Onion Portage	Band 8	*Salix* spp.	11–10

Sources: Mochanov and Fedoseeva 1996c:218; Holmes, VanderHoek, and Dilley 1996:321, table 6-5; Pearson 1999:339; Rasic and Gal 2000:67; Guthrie 1983a:282–283; Kunz, Bever, and Adkins 2003:16–17; Anderson 1988:70

[a] Level not defined

occupants may have sought other alternatives (e.g., dung and lignite [Guthrie 1983a]).

The Beringian industry of the final Lateglacial period is difficult to characterize, and both classification and interpretation of the assemblages have generated debate and controversy (Hamilton and Goebel 1999; Dumond 2001; Holmes 2001). Microblade technology, including diagnostic wedge-shaped microblade cores, appears to be present at Healy Lake, Bluefish Caves, and perhaps the newly reported Nogahabara 1 (Cinq-Mars 1990; Cook 1996; Odess and Rasic 2005). Unlike the earlier microblade industry at Swan Point (see chapter 4), however, uncertainties persist about the stratigraphic context and dating of microblades and microblade cores at each of these sites.

Many of the more firmly dated assemblages between 13.5 and 12.8 cal ka contain various bifacial point forms, and these assemblages lack microblades. A small triangular point was found at Dry Creek, and teardrop-shaped Chindadn forms were found at Walker Road and nearby Moose Creek (Goebel et al. 1996; Pearson 1999). Similar assemblages have been reported from the lowest levels of Chugwater (Tanana Basin) and Little John (southwest Yukon Territory), but they have yet to be firmly dated to greater than 12.8 cal ka (Lively 1996; Easton et al. 2004). At Ushki (central Kamchatka), assemblages containing many stemmed points have been (re)dated to this interval (Goebel, Waters, and Dikova 2003). Similar points are present in the lowest level at Bol'shoi El'gakhchan I (Kir'yak 1996), and lanceolate points with rounded bases may have been produced at Tuluaq Hill in northwestern Alaska (Rasic 2003), but the dating of these assemblages is less clear.

Differences in site function and season of occupation may account for much of the variability among assemblages (e.g., Meltzer 2001: 211–212). Sites such as Ushki and Healy Lake probably were occupied during the warmer months for the hunting of migratory waterfowl on lakes and rivers (Cook 1969). The features at Ushki suggest a protracted settlement, whereas Healy Lake seems to have been occupied for a briefer interval (Goebel and Slobodin 1999). The small occupations in the North Alaska Range—Moose Creek, Owl Ridge, and others—are more likely to have been used in the autumn and early winter, when sheep and elk would have been concentrated in the foothills (Guthrie 1983).

Some archaeologists have suggested, nevertheless, that at least some of the variability in Lateglacial Beringian sites is cultural, and that assemblages in the lowest horizons at Ushki I and V and in the Nenana Valley represent complexes or cultures that may be distinguished from the microblade industry (Dikov 1979; Powers and Hoffecker 1989). Much of the discussion has been influenced by differing views about the relationship between these sites and the early Paleoindian complexes of mid-latitude North America.

In our view, the sites of the final Lateglacial contain a unique industry with limited similarities to industries outside Beringia. Although the presence of microblade technology in some sites suggests a continuing link with adjoining regions of Northeast Asia, other types of assemblages dominate the period between 13.5 and 12.8 cal ka. The latter are not found in Northeast Asia or the North American Plains, and they apparently represent a native Beringian development. While not completely isolated from these neighboring regions by sea or ice at the end of the Lateglacial interstadial, Beringia supported a mesic shrub tundra that differed from the environments of Northeast Asia and the Plains. Much of the uniqueness of the Beringian industry may reflect the peculiar circumstances of life in the shrub tundra: for example, the profusion of small bifacial points might be related to an emphasis on hunting birds and small mammals with throwing darts.

6

The Younger Dryas and the End of Beringia

Cooler climates prevailed across the Northern Hemisphere during the Younger Dryas interval, which lasted for more than a millennium, 12,800–11,300 years ago (12.8–11.3 cal ka). Fossil beetle assemblages in northern Beringia indicate a decline in summer temperatures at this time, while the pollen-spore record in many regions reflects a resurgence of herbaceous tundra vegetation (Elias 2000; Bigelow and Edwards 2001). Minor glacial readvances took place in some mountainous areas (e.g., Ten Brink and Waythomas 1985). Lake levels, which had risen significantly during the Lateglacial interstadial, fell during the Younger Dryas, indicating increased aridity (e.g., Mann et al. 2001:125).

Despite the shift to cooler and drier conditions, the Bering Land Bridge was flooded during this period. Although a seeming paradox, inundation of the Bering Strait during the Younger Dryas presumably reflects the delayed effect of shrinking glaciers on global sea level. Sea level reached the critical threshold of approximately 50 meters below the current level at some point after 12 cal ka (Fairbanks 1989; Elias et al. 1996), and Beringia ceased to exist. Both temperature and moisture increased dramatically as the Younger Dryas ended about 11.5–11.3 cal ka.

Traces of human occupation in Beringia and post-Beringia during Younger Dryas times are more substantial than those of the Lateglacial period. This is particularly true in eastern Beringia or Alaska and the Yukon, where sites are documented in many areas—including the southeast coast of Alaska—that still lack evidence of earlier settlement. Although radiocarbon dating of archaeological remains in this time range is complicated by fluctuations in atmospheric radiocarbon (Mann et al. 2001), soil stratigraphy provides a framework at many sites. Occupation layers are often found below or within a buried tundra soil (i.e., soil-stratigraphic marker) that formed toward the end of and immediately after the Younger Dryas.

Both the inundation of the Bering Land Bridge and the colder and drier climates of the Younger Dryas probably had significant effects on human settlement. Assuming that at least portions of the coast or interior regions of the Land Bridge had been occupied during the Lateglacial, the relatively rapid flooding of central Beringia must have displaced people either westward or eastward, or both (Laughlin 1967). At the same time, the expansion of steppic habitat in response to Younger Dryas climates seems to have generated an increase in the steppe bison population (Matthews 1982:140, fig. 2; Guthrie 1990:285–286).[1] The hunting of bison became an important component of the economy—especially in eastern Beringia—and, in perhaps the strangest development of Beringian human ecology, bison hunters of the North American Plains apparently moved northward to the Arctic during an interval of colder climate.

Beringian archaeology developed a curious mixture of familiar Northeast Asian and North American elements in Younger Dryas times. A microblade technology ultimately derived from Northeast Asia became widespread in both western and eastern Beringia. But lanceolate points—sometimes fluted—that are similar or identical to contemporaneous point types on the Plains also became common in many eastern Beringian or Alaska/Yukon assemblages.

Kamchatka: Ushki

As during the preceding period, the most important archaeological sites in western Beringia are found at Ushki on the Kamchatka Peninsula (figure 6.1). Broad-scale excavation of the Younger Dryas level revealed a large and complex occupation floor that contained many features, artifacts, and associated debris. As during the preceding period, Ushki seems highly atypical and perhaps reflects a richer habitat than other parts of Beringia. Nevertheless, during Younger Dryas times, Ushki exhibits much clearer links with sites in eastern Beringia (e.g., Dikov 1979:71).

The site setting and stratigraphy at Ushki are described in chapter 5. The level dating to the Younger Dryas interval (Layer VI) is found at the localities of both Ushki I and V in the same upward-fining sequence of alluvial silt and silty clay that contains Layer VII (see chapter 5). Layer VI lies in deposits of clay and silt clay approximately 30 centimeters above the latter (Goebel, Waters, and Dikova 2003:502–503). At the time of occupation, the surface of Layer VI apparently rested on the aggrading floodplain of the Kamchatka River. Frost wedge formation is

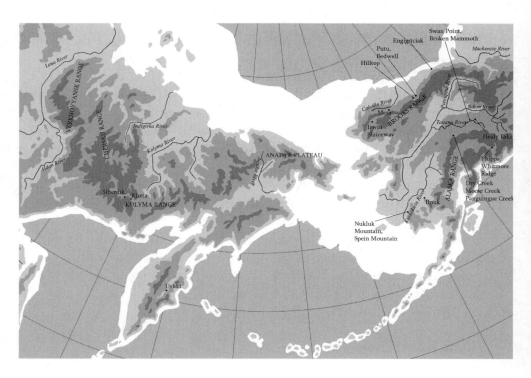

Figure 6.1 Beringia during the Younger Dryas, showing the location of sites mentioned in the text

reported from the eastern portion of the excavated area at Ushki I (Dikov 1977:58).

Floodplain deposition ended after Layer VI was occupied, and a sequence of eolian sand and silt, containing volcanic ash layers and some buried soil horizons of Holocene age, overlies these sediments (Goebel, Waters, and Dikova 2003:502–503). Dikov reported five radiocarbon dates of 12.8–12.0 cal ka (Dikov and Titov 1984:74; Goebel and Slobodin 1999:109–110, table 1),[2] while Goebel, Waters, and Dikova (2003:502–503) recently obtained eight new AMS dates from Ushki I and V, yielding a range of 13.0–11.6 cal ka for Layer VI. Both sets of dates suggest that the occupation falls in the earlier part of the Younger Dryas (table 6.1).

The former houses, hearths, pits, and other features on the Layer VI floor at Ushki I—of which more than 1,800 square meters have been excavated and mapped—reveal a pattern that differs significantly from that of underlying Layer VII (see chapter 5). Traces of multiple structures were uncovered on the occupation floor, and Dikov (1977:52) subdivided them into three categories.[3] To begin, a set of relatively simple features reportedly lie on a lower horizon and antedate the others (Dikov 1990). These are fourteen irregular charcoal stains that range in area between roughly 50 and 150 square meters, most of which contain remains of former hearths—several of them ringed with stones. At least one of these stains is associated with a microblade core and other stone artifacts (Goebel and Slobodin 1999:141), and all of them are interpreted as short-term habitation structures.

The remaining two categories of former structures were found on a slightly higher and younger level of Layer VI (Goebel and Slobodin 1999:140–142, table 3). The first is made up of twelve features ranging in area from 9 to 44 square meters and that exhibit a distinctive "mushroom-like" shape in plan with a rounded semi-subterranean chamber attached to a sunken entrance tunnel (e.g., Dikov 1977:54, fig. 12). The entrance tunnels (which face varying directions except north) would have functioned to minimize the flow of cold air into the living chamber during the winter months (figure 6.2). Eight of the structures contain traces of wooden posts in the form of charred remains of poles or postmolds (in one case with reported diameters of 10–20 centimeters) (Dikov 1977:55), which represent the earliest confirmed use of trees in a Beringian site.

Table 6.1

Calibrated Radiocarbon Dates from Sites in Beringia: Younger Dryas Interval (ca. 12.8–11.3 cal ka)

Site	Level[a]	Material Dated	Lab No.	Uncalibrated Date	Calibrated Age[b]
Ushki 1	Layer VIa	Charcoal	MAG-401	10,360 ± 220	12,122 ± 392
	Layer VI	Charcoal	MO-345	10,360 ± 350	12,039 ± 523
	Layer VI	Carbonized clay	MAG-219	10,760 ± 110	12,741 ± 108
	Layer VI	—	MAG-518	10,790 ± 100	12,778 ± 97
	Layer VIb	Charcoal	MAG-400	10,860 ± 400	12,639 ± 512
	Layer VI	Charcoal	GIN-186	21,000 ± 100	25,150 ± 394
	Layer VI	Charcoal	AA-45720	10,230 ± 70	11,965 ± 186
	Layer VI	Charcoal	AA-45719	10,385 ± 90	12,298 ± 209
	Layer VI	Charcoal	AA-45717	11,130 ± 100	13,034 ± 147
	Layer VI	Charcoal	AA-45713	10,040 ± 130	11,634 ± 261
	Layer VI	Charcoal	AA-45715	10,160 ± 75	11,795 ± 200
	Layer VI	Charcoal	AA-45714	10,800 ± 150	12,774 ± 153
Ushki 5	Level VI	Humates	AA-41387	9,485 ± 275	10,790 ± 365
	Level VI	Charcoal	AA-41386	10,240 ± 75	12,003 ± 212
	Level VI	Charcoal	CAMS-74640	10,460 ± 80	12,380 ± 191
Irwin Sluiceway	—	Hearth feature	Beta-120696	9,550 ± 50	10,911 ± 134
	—	Hearth feature	Beta-134677	10,050 ± 70	11,598 ± 189
	—	Hearth feature	Beta-131336	10,060 ± 80	11,610 ± 190
Mesa	—	Charcoal	DIC-1589	7,620 ± 95	8,440 ± 90
	—	Charcoal	Beta-36805	9,730 ± 80	11,060 ± 150
	—	Charcoal	Beta-50429	10,980 ± 280	12,930 ± 260
	—	Charcoal	Beta-50430	9,945 ± 75	11,440 ± 150
	—	Charcoal	Beta-50428	10,090 ± 85	11,670 ± 210
	—	Charcoal	Beta-52606	10,060 ± 70	11,600 ± 170
	—	Charcoal	Beta-55286	11,660 ± 80	13,550 ± 110
	—	Charcoal	Beta-55285	10,000 ± 80	11,520 ± 170
	—	Charcoal	Beta-55284	9,930 ± 80	11,430 ± 150
	—	Charcoal	Beta-55283	10,240 ± 80	11,990 ± 170
	—	Charcoal	Beta-55282	9,990 ± 80	11,500 ± 170
	—	Charcoal	Beta-57430	11,190 ± 70	13,110 ± 90
	—	Charcoal	Beta-57429	9,900 ± 70	11,380 ± 130
	—	Charcoal	Beta-69900	10,050 ± 90	11,600 ± 200
	—	Charcoal	Beta-69899	9,900 ± 80	11,400 ± 140
	—	Charcoal	Beta-69898	10,070 ± 60	11,610 ± 160
	—	Charcoal	Beta-84650	10,080 ± 50	11,630 ± 150
	—	Charcoal	Beta-84649	9,980 ± 60	11,470 ± 140
	—	Charcoal	Beta-95600	10,230 ± 60	11,950 ± 120

(Table 6.1 continued)

Site	Level[a]	Material Dated	Lab No.	Uncalibrated Date	Calibrated Age[b]
	—	Charcoal	Beta-96070	10,260 ± 110	12,060 ± 260
	—	Charcoal	Beta-96069	10,150 ± 130	11,780 ± 280
	—	Charcoal	Beta-96068	10,080 ± 120	11,670 ± 250
	—	Charcoal	Beta-96067	9,850 ± 150	11,360 ± 260
	—	Charcoal	Beta-96066	10,090 ± 110	11,680 ± 240
	—	Charcoal	Beta-96065	9,810 ± 110	11,250 ± 170
	—	Charcoal	Beta-95914	10,130 ± 60	11,770 ± 160
	—	Charcoal	Beta-95913	10,080 ± 60	11,640 ± 170
	—	Charcoal	Beta-118585	10,130 ± 50	11,780 ± 140
	—	Charcoal	Beta-118584	10,040 ± 50	11,560 ± 140
	—	Charcoal	Beta-118583	10,050 ± 50	11,570 ± 140
	—	Charcoal	Beta-118582	10,100 ± 50	11,690 ± 160
	—	Charcoal	Beta-118581	10,170 ± 50	11,860 ± 110
	—	Charcoal	Beta-119100	10,000 ± 50	11,490 ± 130
	—	Charcoal	Beta-120400	9,740 ± 50	11,170 ± 50
	—	Charcoal	Beta-120399	9,860 ± 50	11,280 ± 50
	—	Charcoal	Beta-120398	9,920 ± 50	11,360 ± 90
	—	Charcoal	Beta-120793	9,800 ± 60	11,220 ± 40
	—	Charcoal	Beta-120397	8,820 ± 230	9,900 ± 280
	—	Charcoal	Beta-125998	10,030 ± 40	11,540 ± 120
	—	Charcoal	Beta-125997	10,080 ± 40	11,630 ± 130
	—	Charcoal	Beta-125996	9,330 ± 40	10,550 ± 60
	—	Charcoal	Beta-125995	9,160 ± 140	10,370 ± 160
	—	Charcoal	Beta-130577	9,780 ± 40	11,210 ± 30
	—	Charcoal	Beta-133354	9,950 ± 60	11,430 ± 140
	—	Charcoal	Beta-133353	10,180 ± 60	11,870 ± 120
	—	Charcoal	GX-26461	12,240 ± 610	14,800 ± 1060
	—	Charcoal	Beta-140199	9,500 ± 190	10,810 ± 270
	—	Charcoal	GX-26567-AMS	9,930 ± 40	11,350 ± 70
	—	Charcoal	Beta-140198	9,480 ± 710	10,920 ± 1010
	—	Charcoal	Beta-142262	10,120 ± 50	11,760 ± 150
	—	Charcoal	Beta-142261	10,080 ± 50	11,630 ± 150
Putu	Zone II	Charcoal	GaK-4941	5,700 ± 190	6,529 ± 201
	Zone II	Soil organics	GaK-4939	6,090 ± 430	6,920 ± 445
	Zone II	Soil organics	WSU-1318	8,450 ± 130	9,409 ± 128
	Below hearth	Charcoal	SI-2382	11,470 ± 500	13,522 ± 608
Bedwell	—	Charcoal	CAMS-11032	10,490 ± 70	12,430 ± 170
Hilltop	—	Charcoal	CAMS-11034	10,360 ± 60	12,290 ± 170
Engigstciak	Buffalo Pit	Bison bone	RIDDL-319	9,400 ± 230	10,701 ± 329
	Buffalo Pit	Bison bone	RIDDL-281	9,770 ± 180	11,186 ± 319
	Buffalo Pit	Bison bone	RIDDL-362	9,870 ± 180	11,396 ± 304
Broken Mammoth	CZ 3A[c]	Soil and charcoal	WSU-4266	9,310 ± 165	10,563 ± 233

(Table 6.1 continued)

Site	Level[a]	Material Dated	Lab No.	Uncalibrated Date	Calibrated Age[b]
	CZ 3	Charcoal	UGA-6256D	9,690 ± 960	11,138 ± 1256
	CZ 3	Charcoal	WSU-4263	10,270 ± 110	12,075 ± 284
	CZ 3	Charcoal	WSU-4019	10,790 ± 230	12,661 ± 300
	CZ 3	Charcoal	CAMS-5357	10,290 ± 70	12,127 ± 228
Swan Point	CZ 3			10,010 ± 90	11,552 ± 197
	CZ 3			10,025 ± 60	11,548 ± 169
	CZ 3	Charcoal	CAMS-4251	10,230 ± 80	11,975 ± 213
Mead	CZ 3	Charcoal	Beta-59118	10,410 ± 80	12,331 ± 194
	CZ 3	Charcoal	Beta-59119	10,460 ± 110	12,364 ± 212
	CZ 3	Charcoal	WSU-4425	10,760 ± 170	12,673 ± 234
Healy Lake	Level 6	Charcoal	Beta-76064	5,110 ± 90	5,852 ± 101
	Level 6	Soil	Beta-76062	7,920 ± 90	8,804 ± 233
	Level 6	Charcoal	GX-2173	10,250 ± 380	11,926 ± 563
	Level 6	Charcoal	CAMS-15920	10,410 ± 60	12,338 ± 180
	Level 6	Charcoal	CAMS-15918	11,100 ± 60	12,995 ± 116
	Level 6	Charcoal	CAMS-15914	11,410 ± 60	13,309 ± 137
	Level 7	Charcoal	GX-2171	8,655 ± 280	9,735 ± 354
	Level 7	Charcoal	GX-2170	8,680 ± 240	9,784 ± 297
	Level 7	Charcoal	CAMS-15919	8,990 ± 60	10,098 ± 116
	Level 7	Charcoal	AU-1	9,245 ± 213	10,494 ± 279
	Level 7	Charcoal	GX-2174	9,895 ± 210	11,448 ± 354
	Level 7	Charcoal	SI-737	10,150 ± 210	11,854 ± 411
	Level 7	Charcoal	CAMS-15917	10,290 ± 60	12,130 ± 213
	Level 7	Plant	CAMS-16523	11,550 ± 50	13,442 ± 111
Spein Mountain	—	Charcoal	CAMS-8281	10,050 ± 90	11,686 ± 336
Dry Creek	Component II	Charcoal	AA-11730	8,915 ± 70	10,035 ± 119
	Component II	Charcoal	AA-11727	10,060 ± 75	11,618 ± 201
	Component II	Charcoal	AA-11728	10,615 ± 100	12,522 ± 173
	Component II	Charcoal	SI-1561	10,690 ± 250	12,508 ± 344
Moose Creek	Component II	Charcoal	Beta-106040	10,500 ± 60	12,432 ± 178
Panguingue Creek	Component I	Carbonized sediment	AA-1687	8,170 ± 120	9,135 ± 168
	Component I	Charcoal	GX-17457	9,836 ± 62	11,265 ± 48
	Component I	Charcoal	AA-1686	10,180 ± 130	11,850 ± 309
Phipps	Occupation layer	Charcoal	UGa-927	8,155 ± 265	9,056 ± 330
	Occupation layer	Charcoal?	UGa-572	10,150 ± 280	11,857 ± 480
	Occupation layer	Charcoal?	CAMS-7659	10,230 ± 270	11,965 ± 186
Whitmore Ridge	Component I	Charcoal?	CAMS-6406	9,890 ± 70	11,362 ± 108
	Component I	Charcoal?	CAMS-8300	9,600 ± 140	10,932 ± 196

(Table 6.1 continued)

Site	Level[a]	Material Dated	Lab No.	Uncalibrated Date	Calibrated Age[b]
	Component I	Charcoal?	CAMS-11255	9,830 ± 60	11,257 ± 43
	Component I	Charcoal?	CAMS-16834	10,270 ± 70	12,083 ± 228
	Component I	Charcoal?	CAMS-16833	10,630 ± 60	12,612 ± 92

SOURCES: Goebel and Slobodin 1999:109–110, table 1; Goebel, Waters, and Dikova 2003:503, fig. 3; Rasic 2003:24, table 1; Kunz, Bever, and Adkins 2003:20–21; Alexander 1987:36; Reanier 1995:40–41; Cinq-Mars et al. 1991; Holmes 2001:158, fig. 2; C. E. Holmes, pers. comm, 2006; Cook 1996:327, table 6-8; Erlandson et al. 1991; Ackerman 2001:91; Bigelow and Powers 2001: 177–179, table 3; Pearson 1999:340, table 2; West, Robinson, and Curran 1996:384, table 8-1; West, Robinson, and West 1996:394, table 8-3

  [a] Dash indicates level not defined; blank space indicates data not available
  [b] Dates calibrated with the Cologne Radiocarbon Calibration Programme CalPal-SFCP-2005 glacial calibration curve (September 2005) (www.calpal.de)
  [c] CZ, Cultural Zone

Each of the twelve semi-subterranean dwellings contained a centrally located hearth lined with stones. Some of the hearths are described as deep (more than 20 centimeters in thickness) and structurally complex with multiple layers of charcoal, ash, and bone (Dikov 1977:55). In one dwelling, the hearth was surrounded by alternating layers of ash and sterile loam, suggesting repeated use over time. Small pits were found in several structures. One of them contained a cache of microblade core performs, and another yielded grinding stones and other artifacts (Goebel and Slobodin 1999:140). Pits with burials were also found in three structures. Two contained traces of humans (in one case the flexed burial of a child associated with red ocher and artifacts [Dikov 1996:245–247]), and the third yielded traces of a domesticated dog—also flexed and associated with red ocher and artifacts (Dikov 1979:57–60) (figure 6.3).[4]

In 2000, Goebel, Waters, and Dikova (2003:503) excavated a similar semi-subterranean dwelling from Layer VI of Ushki V. The former structure measured 5 meters in diameter with a floor excavated 30 centimeters below the ground surface at the time of occupation. A narrow entrance tunnel—oriented to the northwest—was attached to the main living chamber. A centrally located stone-lined hearth (80 centimeters in diameter) and traces of one wooden post (posthole filled with debris) were found in the chamber floor. The hearth contained thousands of small, burned bone fragments, including those of fish and birds, and

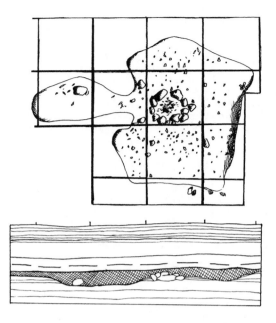

Figure 6.2 Traces of a small house with sunken entrance tunnel mapped by N. N. Dikov on Layer VI at Ushki I (after Dikov 1977:54, fig. 12)

was surrounded by numerous artifacts—microblades and microblade cores, bifaces, and others. More artifacts were found around the perimeter of the living chamber.

Dikov (1977:52) also described another category of former structures on the Layer VI occupation floor at Ushki I. These are circular or irregular in plan, up to 27 square meters in area, and confined to the surface of the occupation layer (i.e., not semi-subterranean). None of the fourteen features in this category possess traces of an entrance tunnel. Half of them occupy an area of less than 10 square meters and may be too small to represent a former dwelling. Most of them, however, contain a centrally located hearth ringed with stones (Goebel and Slobodin 1999:141). No pits or burials were found in this category of features.

The semi-subterranean houses with entrance tunnels may represent winter dwellings (Goebel and Slobodin 1999:135), although the fish and bird bones reported from some of them suggest warm-season use. In any case, their excavated floors and entrances, wooden post construction, pits and burials, and complex hearths all indicate extended

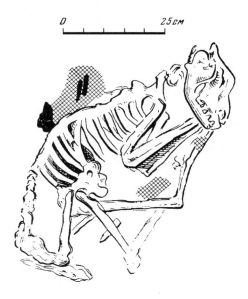

Figure 6.3 Burial containing the remains of a domesticated dog in Layer VI at Ushki I (after Dikov 1979:60, fig. 15)

occupation. The surface features, including those on the older surface underlying the main level, might be traces of more ephemeral structures such as tents occupied during the summer.

No systematic or detailed description of the faunal remains has been published, but it is apparent from scattered references in the literature that a wide spectrum of vertebrates is represented. As already noted, fish and bird bones are reported from some of the hearths (Dikov 1996:245; Goebel, Waters, and Dikova 2003:503), and these are thought to include salmon and duck (Dikov 1977, 1990). Traces of domesticated dog (*Canis familiaris*) are represented in one of the burial pits, and more than 100 incisors of lemming (*Lemmus* sp.) were found in the burial pit containing the child (Dikov 1979). Other vertebrates are listed in table 6.2. The presence of steppe bison may be significant, because this taxon is especially common in other parts of Beringia during the Younger Dryas interval, and may reflect increased steppic habitat at the time.

The artifacts recovered from Layer VI are—like the features from this level—strikingly different from those of the underlying layer. The stone artifact assemblages (from both Ushki I and V) are diverse

Table 6.2

Large Mammal Remains Identified from Layer VI at Ushki

Taxon	Skeletal Parts[a]
Equus caballus (Horse)	
Bison priscus (Steppe bison)	Teeth, scapula[b]
Ovis nivicola (Mountain sheep)	
Alces alces (Moose)	Predominantly teeth

SOURCES: Dikov 1979:57; Vereshchagin 1979
 [a] Blank space indicates data not available
 [b] Other skeletal parts found, but no specific information given

and contain wedge-shaped microblade cores, microblade-core parts (i.e., wedge-shaped core tablets), and microblades. Other stone implements include retouched blades, bifacial points, larger bifaces, end-scrapers, side-scrapers, burins, chopping tools, and anvil stones (Dikov 1977, 1979; Goebel, Waters, and Dikova 2003:504–505). At least one of the bifacial points exhibits a square base (Goebel and Slobodin 1999:136, fig. 17i). Several stone pendants and at least one stone bead were recovered, as well as three sandstone fragments with incised pits or lines (Dikov 1979:60; Goebel, Waters, and Dikova 2003:504) (figure 6.4).

Two nonstone artifacts were preserved. Both represent paddle-shaped objects of bone, the function(s) of which remain unclear. One is a complete paddle-shaped implement manufactured from a bison scapula—measuring 32 centimeters in length—and the other appears to be a fragment of a similar type of implement, possibly fashioned from mammoth bone (Dikov 1977:56, 1979:63). They may have been used as small boat paddles, snow knives, or flat shovels, and they serve as a reminder that much of the technology of the Ushki residents—bone, antler, hide, wood, and other generally perishable materials—is invisible.

Ushki Layer VI represents one of the most important Younger Dryas sites in Beringia and is the only one to provide traces of living structures, including relatively long-term structures (winter houses?). As in the case of the underlying occupation floor (Layer VII), the Younger Dryas level at Ushki may reflect a more productive habitat

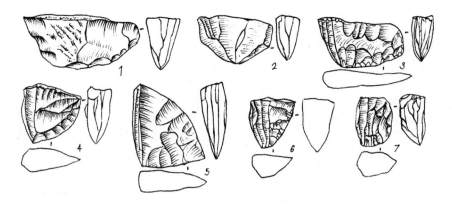

Figure 6.4 Microblade cores from Layer VI at Ushki, which contains an assemblage of wedge-shaped cores, microblades, and other items similar to those of the Denali complex of interior Alaska (after Dikov 1979:61, fig. 16)

than other, higher-latitude parts of Beringia, with a relatively high carrying capacity. There are no other known sites in Beringia of comparable size and complexity.

The artifacts and faunal remains at Ushki are similar, nevertheless, to those found elsewhere in both western and eastern Beringia. Microblade technology is extremely common in sites of the Younger Dryas era, in contrast to the preceding half millennium, and is particularly well dated to this interval in eastern Beringia (West 1981; Hamilton and Goebel 1999). The appearance of steppe bison in Ushki Layer VI also offers a parallel to a pattern found in eastern Beringia during the Younger Dryas.

Western Beringia: Other Sites?

Several other sites in western Beringia contain remains that may date to the Younger Dryas interval, but their dating is problematic. The sites may actually date to the warmer period that followed after 11.5–11 cal ka. Outside Kamchatka, sites are not firmly documented in western Beringia until the warmer period (Pitul'ko 2001:269–270), raising the question—given the number of earlier occupations found in Alaska and the Yukon—of whether or not the population of western Beringia remained comparatively low at this time. Alternatively, the scarcity of

sites might reflect the limited scope of modern settlement and archaeological survey in the region.[5]

The best known of western Beringian sites that may date to the
Younger Dryas is Siberdik, which is located at latitude 61° 36′ N in the
Upper Kolyma Basin near the mouth of a major tributary (Detrin
River). The site lies in the northern uplands of the Kolyma Range at an
elevation of roughly 415 meters above sea level (T. Goebel, pers. comm.,
2005). It was discovered in 1971, excavated by Dikov in 1971–1975, and
now is submerged beneath the waters of the Kolyma Reservoir. Siberdik was formerly situated on a 14-meter high promontory along the east
side of the river and adjacent to a small side-valley stream (Dikov
1977:213–214).

Three occupation layers are buried in a sequence of sand, sandy
loam, and loam deposits that overlie coarse alluvium capping the bedrock terrace. Given the elevation of the terrace above the pre-reservoir
level of the Detrin River, the sediment containing the archaeological
remains is probably of eolian rather than alluvial origin. The uppermost occupation layer at Siberdik lies in the modern soil and is dated
to less than a thousand years ago. A middle layer is buried at a depth of
50–80 centimeters below the surface and yielded five radiocarbon dates
of 7.5–5 cal ka. The lowermost layer (Cultural Layer III) lies 90–120
centimeters below the surface in a partly eroded "peaty horizon" and
has been dated to 11–8 cal ka (Dikov 1977:213–218; Goebel and
Slobodin 1999:114).[6]

Following his practice at other sites, Dikov (1977:213) excavated a
large area at Siberdik (more than 800 square meters), uncovering an
occupation floor on Layer III with former hearths and concentrations
of artifacts and other debris. Also present on this level was a complex
arrangement of polygonal frost cracks, and in some cases, artifacts were
found redeposited in the wedges (Dikov 1977:216–217, fig. 171). The
hearths contained charred stones (possibly indicating stone rings similar to those at Ushki) and burned and calcined bones. According to
Dikov (1977:218), possible traces of a human burial with red ocher are
present. Faunal preservation was poor, but a few identified taxa are
reported and include horse teeth, deer antler, and the scapula of an
unidentified herbivore (Dikov 1977:218–221).

The artifact assemblage from Cultural Layer III was composed of a
wedge-shaped microblade core, microblades, burins (including at least

two polyfaceted specimens), side-scrapers, end-scrapers (on large blades), "knives" on blades, and bifacial points. The assemblage also contains a number of large cobble tools, hammer stones, and anvil stones (Dikov 1977:218, 1979:92–96). Overall, it is similar to the Ushki Layer VI assemblage (and more so than to younger artifact assemblages of the Kolyma Basin [e.g., Slobodin 2001:35]), and although the mean date for Cultural Layer III is significantly less than 11 cal ka, it is the most likely occupation of Younger Dryas age in the region. The frost-wedges apparently associated with the occupation indicate a cold episode.

Another site in the Upper Kolyma Basin that may date to the Younger Dryas is Kheta, located on a different tributary (Kheta River) at approximately 800 meters above sea level. The site occupies a bedrock surface (or third terrace level) 15 meters above the modern river. Kheta was discovered and investigated by S. B. Slobodin in 1990–1992 (King and Slobodin 1994; Slobodin and King 1996).

A relatively shallow sequence of unconsolidated deposits—no more than 25 centimeters in thickness—overlies the schist bedrock (younger deposits apparently were removed locally by recent construction activity [Goebel and Slobodin 1999:112]). Beneath a thin surface organic layer lies a volcanic tephra unit that is tentatively assigned to the Elikchan Tephra (ca. 9.4 cal ka). Eolian sand horizons exhibiting traces of soil formation underlie the tephra.[7] More than 500 artifacts were found at the contact of the sand horizons and underlying coarse gravel unit (Slobodin and King 1996:240–241). Among the artifacts are a wedge-shaped microblade core, ski spall, microblades, burin, end-scrapers, perforator, and several bifaces and biface fragments. Also present is a polished stone bead and small pendant (Slobodin and King 1996:242, fig. 4-15). The assemblage has been compared with those from Dyuktai sites in the Lena Basin.

Northern Alaska and Yukon

The 1947 discovery of a fluted projectile point on a ridge overlooking the upper Utukok River in Alaska sparked immediate interest in possible connections between Beringia and early Paleoindian sites of mid-latitude North America (Thompson 1948).[8] The Utukok River drains the northwestern Brooks Range and empties into the Arctic Ocean. The point, which measures 5.5 centimeters in length and exhibits multiple flutes, bears some resemblance to the Folsom point type (Thompson 1948:62, fig. 9). The latter is associated with early bison hunting on the High Plains and currently is dated to 12.8–11.8 cal ka (Holliday 2000:264–276) (figure 6.5).

In the years after the find on the Utukok River, more fluted points were discovered in northern Alaska and the Yukon. In 1950, two such points—thought to resemble Folsom and Plainview, respectively—were encountered on a knoll along the Kugururok River in the northwestern foothills of the Brooks Range (Solecki 1996:519–520). In 1955, Richard S. MacNeish excavated a fragmentary bifacial point that had been fluted on one surface from the site of Engigstciak, located near the mouth of the Firth River on the Arctic Ocean coast of the Yukon (MacNeish 1956:96). Subsequent fluted point finds were made at Batza Téna, Girls Hill, Putu-Bedwell, Mesa, and other localities on the north and

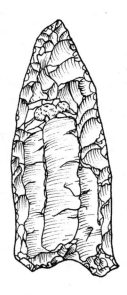

Figure 6.5 Fluted point recovered from a surface context on the Utukok River in northern Alaska in 1947; like many of the fluted points found in eastern Beringia, it exhibits multiple fluting scars (drawn from a photograph in Thompson 1948:62, fig. 9)

south slopes of the Brooks Range and in northern Yukon (Clark 1984).

In addition to the fluted points, unfluted lanceolate points and point fragments were recovered from many of the same sites, as well as other localities in northeastern Beringia (Hamilton and Goebel 1999:172–178). Many of these lanceolate points were compared with late Paleoindian or Plano tradition point types of the Plains, such as those at Agate Basin, Plainview, and Hell Gap (e.g., MacNeish 1956:96–97; Reanier 1996:507; Kunz, Bever, and Adkins 2003:29). The late Paleoindian complexes are also associated with bison hunting on the High Plains and currently date to as early as 12.8 cal ka— overlapping chronologically with Folsom (Holliday 2000:264–276). Similar types of lanceolate points are known from other parts of Beringia, including the Tanana Basin, where they were reported as early as 1939 (Rainey 1940).

Both the fluted and unfluted lanceolate points in northeastern Beringia were found in surface or shallow sedimentary contexts that proved difficult to date (Clark 1984; Hamilton and Goebel 1999:180–182). Although similar to Folsom, few if any of the fluted forms may be considered classic examples of this type, and most exhibit multiple flutes (e.g., Dixon 1999:187). In 1975, a radiocarbon date of 13.5 cal ka was reported on charcoal from a former hearth associated with a fluted point from Putu (Alexander 1987:42), suggesting that the latter might be contemporaneous with the oldest fluted points known in mid-latitude North America. Reanalysis of the laboratory records revealed, however, that the date was obtained on charcoal recovered from below the occupation layer (Reanier 1996:509). Several dates on samples from the hearth at Putu indicate an age of ca. 9.6–9.4 cal ka (Hamilton and Goebel 1999:174–175).

New radiocarbon dates from several sites now suggest that at least some of the fluted points and many of the unfluted lanceolate points from northeastern Beringia date to the Younger Dryas interval (Mann et al. 2001:123, fig. 2). Most significant is the large sample of dates from the Mesa site in the northern Brooks Range, which exhibits a cluster between 12 and 11 cal ka (Kunz, Bever, and Adkins 2003:19–23). Dates in the same time range are also reported from Engigstciak, Bedwell, and Hilltop (Cinq-Mars et al. 1991; Reanier 1995, 1996:510). Because these point types seem to have been developed earlier in mid-latitude North America, the Beringian specimens are thought to represent a northward movement of people or their artifacts from the High Plains (e.g., MacNeish 1963:101–102; Dixon 1999:187–188; Bever 2001). The temporal relationship between the fluted and unfluted points in Beringia remains unclear but may parallel that of the High Plains (i.e., fluted forms antedate but overlap with unfluted types).

Few faunal remains are preserved in the northeastern Beringian sites, but renewed investigation at Engigstciak produced bones of steppe bison with butchering marks dated to 11.4–10.7 cal ka (Cinq-Mars et al. 1991). An emphasis on bison hunting also is suggested for fluted and lanceolate point sites in the northern Brooks Range (e.g., Mann et al. 2001:132–133), and bison are common in occupations of comparable age in other parts of Beringia, especially central Alaska (e.g., Dry Creek [Guthrie 1983a], Broken Mammoth [Yesner 2001]). The apparent increase in bison numbers presumably reflects the cooler and drier conditions of the Younger Dryas and the expansion of steppic vegetation. The trend seems to have brought High Plains hunters into eastern Beringia after 12.8 cal ka, creating the first demonstrable cultural link between the two regions.

Assemblages containing wedge-shaped microblade cores and microblades, which are common in other parts of Beringia during the Younger Dryas, may also be present in the Brooks Range at this time, but their dating is uncertain. In 1950, microblade cores were recovered from a surface locality of unknown age on the Kukpowruk River near the Arctic Ocean coast (Solecki 1996:517–518). Microblades and other artifacts at Locality 1 of the Gallagher Flint Station (Sagavanirktok River) yielded a radiocarbon date in this time range (Dixon 1975), but according to Ferguson (1997), the relationship of the dated material to the artifacts is problematic.

Concentrations of microblades were found at Mesa, but Kunz, Bever, and Adkins (2003:34–38) believe that they represent a separate and younger occupation. Some microblades are reported from the assemblage containing lanceolate points at Engigstciak (Cinq-Mars et al. 1991:35). Microblades have also been found in lanceolate point assemblages from the Noatak River region (assigned to the Sluiceway complex), but here again the relation between the dated materials and artifacts is unclear (Rasic 2003:24).

WESTERN BROOKS RANGE

In 1993, Robert Gal and Thomas Hamilton discovered a site that later was named Irwin Sluiceway. The site is located on a tributary of the Noatak River system (Anisak River) in the western Brooks Range. Traces of a hearth, discovered in 1998, yielded radiocarbon dates of 11.6–10.9 cal ka (Rasic 2003:24). Among the artifacts were a number of bifacial lanceolate points exhibiting impact scars and basal edge-grinding. Along with as many as eighteen other sites in the Noatak River Basin (including NR-5 and MIS-495), this assemblage has been assigned to the Sluiceway complex (Rasic and Gal 2000; Rasic 2003). As noted above, microblades are present in at least some of these sites.

MESA

The Mesa site rests on a 60-meter high block of extruded igneous rock that overlooks the northern slope of the Brooks Range at latitude 68°N (figure 6.6). The location provides an unobstructed view of the

Figure 6.6 Topographic setting of the Mesa site (photograph courtesy of M. L. Kunz)

surrounding landscape and more than 3,000 square meters of suitable occupation area (Kunz, Bever, and Adkins 2003:7–11). Discovered in 1978 by Michael Kunz and Dale Slaughter, the Mesa site has been subject to thirteen seasons of excavation and exposure of a large area (roughly 300 square meters).

The gabbro bedrock is covered with a thin layer of windblown silt or loess that varies in thickness from 5 to 35 centimeters (Kunz and Reanier 1996:497–499). Although there is little natural or cultural stratigraphy at the site, most artifacts and features appear to represent a tightly constrained period of occupation, and some of them lie on or near the base of the loess (Kunz, Bever, and Adkins 2003:12–14). More than fifty radiocarbon dates now have been obtained on charcoal samples recovered from twenty-eight former hearths. The dates yield a relatively narrow peak between 12 and 11 cal ka (Kunz, Bever, and Adkins 2003:19–23). The Mesa site was occupied most intensively during the later Younger Dryas (Mann et al. 2001:122–125).

A total of forty former hearths have been excavated at Mesa, and—aside from the concentrations of lithic debris—they represent the only features at the site. The former hearths are small and typically less than 12 centimeters thick, and many were found in shallow depressions intruding into the bedrock (Kunz, Bever, and Adkins 2003:15). Most hearths contained small fragments of charcoal—apparently representing willow, poplar, and alder—and occasional pieces of burned bone, but one hearth at the northeast end of the occupation area ("East Ridge Locality") yielded a large quantity of burned bone (Kunz, Bever, and Adkins 2003:16–17).

More than 120,000 stone artifacts have been recovered from the Mesa site and fewer than 1 percent represent worked implements. The formal tools are dominated by bifacial projectile points or point fragments (more than 150 examples) (Kunz, Bever, and Adkins 2003:27). The points are described as lanceolate in form—diamond-shaped in cross section—with slightly tapering sides and flat or concave bases. Some specimens exhibit basal fluting. Edges were heavily ground along roughly two-thirds of their total length, suggesting that the points were "set deeply in the haft" (Kunz and Reanier 1996:500). Other worked items include some larger bifaces of varying shape, scraping tools (including end-scrapers), and gravers. The gravers were made on flakes and exhibit one or more spurs, and most have been edge-retouched—they represent composite tools that had multiple functions (Kunz, Bever, and Adkins 2003:29–32) (figure 6.7).

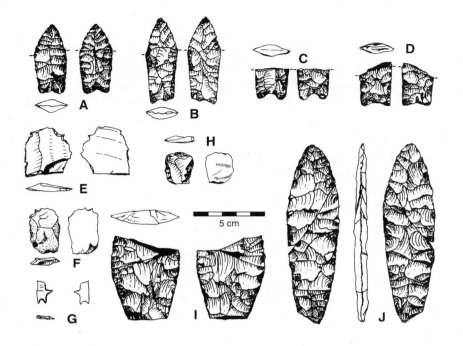

Figure 6.7 Artifacts from the Mesa site, including lanceolate points and spurred gravers (reprinted with permission from Kunz and Reanier 1994:661, fig. 2. Copyright 1994 AAAS)

Also present among the Mesa site assemblage were a wedge-shaped microblade core, microblade core parts, and at least 130 microblades or microblade fragments. Associated with these items were four bifaces—one of which is fluted—as well as some bifacial trimming flakes (Kunz, Bever, and Adkins 2003:34–38). Most of these artifacts were found concentrated in the southwest area of the occupation ("Locality A") and are believed to represent a separate assemblage and younger component (Kunz, Bever, and Adkins 2003:38).[9]

The projectile points from Mesa are similar to points found in later Paleoindian assemblages of the Plains—especially Agate Basin and Hell Gap—and Kunz and Reanier (1994:660) accordingly assigned it to a new Paleoindian complex, the Mesa complex. In addition to the projectile points, other artifact types in the assemblage, including the larger bifaces, scrapers, and gravers, are also diagnostic of Paleoindian complexes on the Plains (Kunz and Reanier 1996:503). Other similar assemblages of comparable age from the Brooks Range have been added to the Mesa complex (Bever 2001:100–108).

Although no identifiable faunal remains were recovered from Mesa, it has been suggested that the occupants of the site were hunters of steppe bison. This is based on the association of the similar late Paleo-indian industry of the Plains with bison hunting, as well as on the documented increase in bison numbers in Alaska—and correspond-ingly low numbers of caribou—during the Younger Dryas interval (Kunz, Bever, and Adkins 2003:48–52). Similar types of projectile points are associated with cut-marked bones of steppe bison in a slightly younger occupation at Engigstciak in the northern Yukon (Cinq-Mars et al. 1991).

PUTU-BEDWELL AND HILLTOP

Several other sites located in the northern Brooks Range have produced artifacts similar to those of Mesa. They are found in the eastern part of the range, near the confluence of the Sagavanirktok and Atigun Rivers, roughly 250 kilometers from the Mesa site and in similar topographic settings. The localities of Putu and Bedwell (100 meters apart but originally reported as separate sites) occupy a bed-rock hill that overlooks the Sagavanirktok Valley. They were discovered in 1970 and initially investigated by Herbert L. Alexander (1987), and they have been researched more recently by Richard E. Reanier (1995, 1996).

At the Putu locality, which rests on a bench 30 meters below the top of the hill, a 25–50- centimeter thick layer of eolian sediment over-lies the weathered shale bedrock (Alexander 1987:7–9). Most of the artifacts were found in a tan loess unit below the modern soil, along with two former hearths. Four fluted point fragments—three with multiple fluting scars—have been reported from Putu (Alexander 1987:12–14; Reanier 1996:507). The assemblage also contains four lan-ceolate projectile points or point fragments with slightly tapering sides and flat or convex bases that are similar to the points from Mesa. Other items include large bifaces, burins, scrapers, and gravers (Alexander 1987:20–24).

Six radiocarbon dates have been obtained on samples (mostly charcoal) from various depths at the Putu locality, and they range from the late Holocene to the terminal Pleistocene (Reanier 1996:509, table 11-2). The date of 13.5 cal ka was originally reported from one of the hearths in the loess unit (Alexander 1987:42). This early date in apparent association with fluted points generated considerable interest in Putu as a possible early Paleoindian site above the Arctic Circle. According to Reanier (1996:509), however, laboratory records indicate that the sample was derived from a feature in the weathered bedrock zone below the occupation level. Two dates on samples thought to be more closely associated with the latter suggest an early Holocene age (10–9 cal ka).

On the crest of the same hill at the Bedwell locality, artifacts were excavated from shallow loess deposits similar to those at Putu. The assemblage contains lanceolate projectile points with gently tapering sides and flat or convex bases. One point base is concave, and the flaking methods are generally similar to those deduced from the Mesa points (Reanier 1996:510). A single radiocarbon date on charcoal thought to be associated with some of the points is 12.4 cal ka—suggesting an occupation at roughly the same time as Mesa (Reanier 1995, 1996; Mann et al. 2001:122–124).

The Hilltop site is located on a bedrock hill that overlooks the Atigun River Gorge. The artifact assemblage from this site contains lanceolate projectile points with gently sloping sides and concave or slightly convex bases, large bifaces, scrapers, and gravers (Bever 2001:104–105). Associated charcoal yielded an AMS radiocarbon date of 12.3 cal ka (Reanier 1995:41).

ENGIGSTCIAK

Engigstciak, which means "new mountain," is found in a topographic setting similar to that of the sites in the Brooks Range. The site occupies a bedrock knoll near the mouth of the Firth River in northern Yukon and was discovered by the late Richard S. MacNeish in 1954. At latitude 69°N and 25 kilometers from the shore of the Arctic Ocean, Engigstciak remains one of the northernmost sites of terminal Pleistocene or earliest Holocene age—surpassed only by Berelekh and Zhokhov Island. MacNeish investigated Engigstciak between 1955 and 1958, excavating several areas of varying size and roughly 190 test pits across the crest and slopes of the knoll (Mackay, Mathews, and MacNeish 1961:27–29).

The Engigstciak knoll is composed of shale and covered with a thin layer of Quaternary sediment. On the crest of the knoll, the modern vegetation mat overlies deposits of marine clay separated by an organic-rich silt or clay band (Mackay, Mathews, and MacNeish 1961:29–33). The sequence has been heavily disturbed by cryoturbation and gelifluction, and the stratigraphic relationships of the artifacts are problematic. On the southern slope, however, an area of relatively undisturbed stratigraphy was encountered. In this area—designated the "Buffalo Pit"—bedded humic sands underlie the vegetation mat. Near the base of this layer lie artifacts of the Arctic Small Tool tradition (ASTt) and remains of large and small mammals. Below the humic sands lie pale brown and yellowish sands that yielded lanceolate points and bison bones (Mackay, Mathews, and MacNeish 1961:36).

MacNeish (1956:92) defined no fewer than nine archaeological complexes at Engigstciak, including two early complexes below the ASTt level that attracted interest because little was known about the oldest arctic industries at the time. The lowest cultural level—containing scrapers and large planoconvex scraping tools—was designated the British Mountain complex. The overlying level, which had produced the lanceolate points and bison bones, was assigned to the Flint Creek complex. MacNeish (1956:96) compared the lanceolate points with Angostura and Plainview points of the Plains and also noted the presence of a fluted point fragment. Other items included scrapers, burins, and a microblade fragment.

Interest in the early complexes at Engigstciak faded in the face of concerns regarding the disturbed stratigraphy and a lack of radiocarbon dates (Cinq-Mars et al. 1991:34–35). A few years ago, Jacques Cinq-Mars and colleagues examined the large mammal remains collected from the Buffalo Pit and obtained radiocarbon dates on three bones. More than 300 bison bones (*Bison priscus*) had been recovered from the Flint Creek level, along with remains of caribou and musk ox. Tool cut-marks were observed on several of the bison bones. The dated samples yielded ages of 11.4–10.7 cal ka (Cinq-Mars et al. 1991). Although Engigstciak may be somewhat younger than the sites in the Brooks Range (and postdate the Younger Dryas), the association between the generally similar lanceolate points and steppe bison is significant.

Tanana River Valley

As during the Lateglacial period, there are several major occupations in the Tanana Valley that date to the Younger Dryas or its immediate aftermath. Near the mouth of Shaw Creek, the sites of Broken Mammoth, Swan Point, and Mead all contain occupation levels that date to the Younger Dryas. Although it has been suggested that these levels actually postdate the Younger Dryas (Yesner 1995), radiocarbon dates indicate that they fall in the range of 12.7–11.5 cal ka (Holmes 2001:158; C. E. Holmes, pers. comm., 2006). The preservation of faunal remains is excellent—especially at Broken Mammoth—and these sites will contribute significantly to knowledge about the economy and diet of people in eastern Beringia during this interval (Yesner 2001:319–320).

The broad range of dates and continuous sequence of occupation layers from the lower levels at Healy Lake suggests that people were here during the Younger Dryas as well. Some of the assemblages recovered from these sites are of ambiguous affiliation, but both a microblade industry similar to that of Ushki Layer VI (known locally as the Denali complex) and lanceolate points similar to those of northern Alaska and the Plains (Mesa complex) are present.

Several well-known sites in the Tanana Valley probably were occupied immediately after the Younger Dryas event (i.e., after 11.3 cal ka). Southwest of Healy Lake and along the lower course of the Gerstle River, another bedrock knoll location (Gerstle River) seems to have been visited for the first time at ca. 11.2 cal ka (Potter 2002:89–90). Occupations at Chugwater and Campus—located downriver near Fairbanks—also appear to have occurred after the Younger Dryas, when climates were significantly warmer (Lively 1996:308–309; Pearson and Powers 2001).

BROKEN MAMMOTH

Located on a bedrock bluff at the confluence of Shaw Creek and the Tanana River, Broken Mammoth contains occupation levels dating to both the Lateglacial and Younger Dryas buried near the base of deep eolian deposits of sand and silt (see chapter 4). The second lowest horizon ("Cultural Zone 3") has yielded four radiocarbon dates that provide an age of 12.6–11.1 cal ka (Holmes 1996:315). The artifacts, features, and faunal remains in this horizon are associated with a buried soil made up of two to three undulating organic-rich bands ("middle paleosol complex") in a thick bed of pale to light-yellowish brown loess ("lower loess") (Holmes 1996:313, table 6-3).

The artifact assemblage from Cultural Zone 3 contains the base of a concave bifacial projectile point, a complete subtriangular bifacial point, a larger biface, retouched flakes, hammer stones, and a large quantity of small waste flakes. Also present are anvil stones of schist. Rhyolite and chert otherwise predominate among raw materials. A small but complete eyed needle of bone was recovered in association with a dated former hearth (Holmes 1996:317) (figure 6.8). At least one other former hearth was uncovered on this level, and both were associated with hearthstones, artifacts, and faunal debris.

Faunal remains from Cultural Zone 3 are listed in table 6.3 and provide more evidence of a broad-based diet but exhibit some important differences from those at the lowest level (see chapter 4). Most striking is the predominance of steppe bison among the large mammals, reflecting either the increased numbers of bison during the Younger Dryas or a specific focus on this taxon by the site occupants,

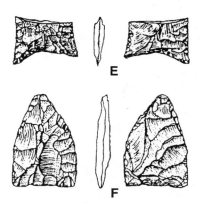

Figure 6.8 Artifacts from the middle level at Broken Mammoth (Mesa complex?), which is dated to the Younger Dryas interval (from Holmes 2001:166, fig. 9; *Arctic Anthropology Journal,* vol. 38, no. 2. Copyright 2001. Reprinted by permission of University of Wisconsin Press)

Table 6.3
Vertebrate Remains from Cultural Zone 3 of Broken Mammoth[a]

	No. of Specimens	Percentage of Specimens
Bison priscus (Steppe bison)	133	22
Cervus elaphus (Elk)	87	14
Rangifer tarandus (Caribou)	6	1
Ovis dalli (Mountain sheep)	11	2
Ursus sp. (Bear)	1	<1
Canis sp. (Wolf)	1	<1
Alopex lagopus (Arctic fox)	13	2
Lepus sp. (Hare)	33	5
Marmota flavescens (Hoary marmot)	8	1
Ochotona collaris (Collared pika)	3	<1
Spermophilus parryi (Arctic ground squirrel)	298	49
Sorex arcticus (Arctic shrew)	9	1
Total Identified Mammals	603	100
Cygnus columbianus (Tundra swan)	41	40
Branta canadensis (Canada goose)	2	2
Anser albifrons (White-fronted goose)	12	12
Chen hyperborea (Snow goose)	5	5
Anas platyrhynchos (Mallard)	6	6
Anas acuta (Pintail)	2	2
Anas strepera (Gadwall)	4	4
Anas americana (Widgeon)	2	2
Anas carolinensis (Green-winged teal)	6	6
Lagopus lagopus (Willow ptarmigan)	23	22
Total Identified Birds	103	100

SOURCES: Yesner 2001:321–322, table 2; D. R. Yesner, pers. comm., 2006

[a] Additional remains include unidentified medium/large mammals (*n* = 639), microtine rodents (*n* = 165), unidentified small mammals (*n* = 108), unidentified mammals (*n* = 203), unidentified birds (*n* = 117), and cycloid/salmonid fish (*n* = 28).

or both. Sectioning of both bison and elk teeth indicate fall mortality for these taxa (Yesner 2001:319).

Although most of the small mammal and bird taxa identified in the underlying Lateglacial occupation are present, there is a significant decrease in the quantity of bird remains (more than 80 percent reduction in Cultural Zone 3). In assemblages of comparable total size, the most abundant waterfowl taxon (tundra swan) declines by more than 90 percent, and there is also a reduction in bones of willow ptarmigan (70 percent). Fish are identified for the first time, however (Yesner 2001:322), offering a parallel with Ushki.

The cultural affiliation of the Cultural Zone 3 artifacts has been something of a puzzle (e.g., Holmes 2001:165). The most clearly defined industry in central Alaska during this interval is the Denali complex (West 1967), but few if any of the diagnostic elements of this complex (described below) are present in Cultural Zone 3 at Broken Mammoth. The concave-based projectile point exhibits similarities to some of the point bases found at Mesa complex sites in the Brooks Range (Reanier 1996:506–508; Kunz, Bever, and Adkins 2003). These observations—in conjunction with the high number of steppe bison remains—suggest that the assemblage might be affiliated with the Mesa complex.

SWAN POINT

Occupying a small bedrock knoll on the northern margin of Shaw Creek Flats, Swan Point contains a compressed version of the eolian stratigraphy at Broken Mammoth (located more than 10 kilometers south) (see chapter 4). At this site, an occupation level dating to the Younger Dryas time range is buried in tan loess 50–55 centimeters below the surface and associated with a paleosol complex (Holmes, VanderHoek, and Dilley 1996:320). Four AMS radiocarbon dates on hearth charcoal indicate an age of 12.4–11.5 cal ka.

Not much information is available on faunal remains, but the artifact assemblage is apparently large and contains microblades, bifacial implements, graver spurs manufactured on broken points, and quartz pebble tools (Holmes, VanderHoek, and Dilley 1996:321–322). The bifacial forms include two triangular points or knives and a small lanceolate point base (with straight sides). As in the case of the contemporaneous assemblage at Broken Mammoth, the affiliation of these artifacts is somewhat problematic. However, the presence of microblades suggests that it is related to the Denali complex, which is well documented in this time range.

HEALY LAKE

Located roughly 60 kilometers upriver from Shaw Creek (see chapter 5), the Healy Lake site most probably contains an occupation of Younger Dryas age, but it is not easily distinguished as a discrete level owing to the more compressed stratigraphy and continuous succession of occupation levels at this site (Cook 1996:325, fig. 6-10). The lower five cultural layers (Levels 6–10) yield a broad range of calibrated radiocarbon dates between 13.4 and 5.9 cal ka, and many of the older dates were obtained on samples from the uppermost levels (Cook 1996:327, table 6-8). Levels 6 and 7 are associated with a weakly buried soil horizon (faint gray A2b soil horizon) in the loess that may have developed toward the end of—or after—the occupations contained in these levels. The character and context of this soil has analogs at other sites in central Alaska (e.g., Broken Mammoth and Dry Creek), and it may represent a widespread marker horizon for the end of the Younger Dryas. On the basis of their stratigraphic position, Levels 6 and 7 may be tentatively assigned to the Younger Dryas and later.

Several former hearths were encountered in Level 6, one of which contains fire-cracked rocks and burned bone fragments (some of which are bird bones) (Cook 1969:240). Former hearths in Level 7 also yielded large quantities of burned bone (Cook 1996:324), while a concentration of decayed mammal bone fragments on this level included an articulated group of carpals or tarsals (Cook 1969:241).

Artifacts from Levels 6 and 7 include at least one wedge-shaped microblade core and eighty-three microblades (which represent 90 percent of the microblades recovered from the lower five levels [Cook 1996:326, table 6-7]). Among other types are bifaces, end-scrapers, and points. Level 6 produced a thin and finely flaked projectile point tip exhibiting straight sides (Cook 1969:184). Although West (1981:139) apparently assigns these artifacts to the Denali complex, Cook (1996:324–325) placed all five of the lower levels at Healy Lake into a separate entity (Chindadn complex) that contains wedge-shaped cores and microblades but differs in other respects from Denali.

Southwest Alaska: Kuskokwim River Basin

Several sites located on tributaries of the Kuskokwim River probably were occupied during the Younger Dryas interval, although their dating remains uncertain. Most of these sites were discovered by Robert E. Ackerman, who conducted surveys of the region in 1979 and 1981 and further investigations during 1992–1993 (Ackerman 1996a, 1996b,

1996c, 1996d). Although the sites in the Kuskokwim drainage lack deep stratigraphy and—with the exception of Cave 1, Lime Hills— contain few faunal remains, they appear to document the presence of both the Denali and Mesa complexes in southwest Alaska (Ackerman 2001).

Spein Mountain is located on a bedrock ridge overlooking the Kisarilik River, roughly 300–360 meters above sea level. Artifacts are buried in a layer of dark brown loess 15–25 centimeters in thickness that overlies weathered basalt (Ackerman 1996b:456–457). Although artifacts were found throughout the loess (and many were collected from surface exposures), most were concentrated at 15 centimeters below the surface. The majority was recovered from areas near the top of the ridge, but several smaller artifact clusters were found on lower slopes (Ackerman 2001:84–85).

The single reported date for Spein Mountain was obtained on small charcoal fragments recovered from a dark lens within a pit feature. The pit feature—which also yielded biface fragments, flaking debris, and burned bone—was discovered in a test excavation unit on the south slope of the ridge (Ackerman 2001:92, fig. 8). The AMS radiocarbon date is 11.7 cal ka. A pollen sample extracted from the same lens reflected predominance of grasses and some woody shrubs, which appears to be consistent with the radiocarbon date (Ackerman 1996b:457).

All of the artifacts recovered from Spein Mountain—including the surface finds—have been lumped into a single assemblage of 4,300 items (Ackerman 1996b:457, table 10-1). Local argillite served as the primary raw material, and projectile points and point fragments were the most common type of tool. The points were lanceolate to ovate in shape with slightly tapering sides and flat to slightly convex bases. Among the fragments, bases were abundant (Ackerman 2001:86). The assemblage also contained a few larger bifaces, scrapers, and gravers (figure 6.9). The debitage included bifacial thinning flakes.

On the basis of the morphology of the lanceolate projectile points and the overall composition of the lithic assemblage, Spein Mountain is assigned to the Mesa complex of the Paleoindian tradition (Ackerman 1996b:460; Bever 2001:105–108). Spein Mountain was the first site outside northern Alaska to be placed in this category, but—as discussed elsewhere in this chapter—there is evidence for the Mesa

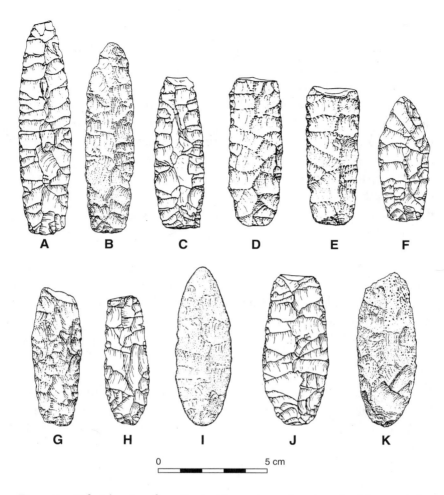

Figure 6.9 Bifacial points from Spein Mountain, assigned to the Mesa complex (from Ackerman 2001:88, fig. 5; *Arctic Anthropology Journal,* vol. 38, no. 2. Copyright 2001. Reprinted by permission of University of Wisconsin Press)

complex in other parts of central and southern Alaska during Younger Dryas time.

About 10 kilometers downstream from Spein Mountain and between the Kisarilik and Fog Rivers, the north slope of Nukluk Mountain yielded an assemblage of wedge-shaped microblade cores, microblades, one multifaceted burin, and bifacial point fragments (Ackerman 1996c:461, table 10-3). The roughly 1,500 artifacts were recovered from the weathered bedrock surface and shallow loess deposits, and they remain undated. Nevertheless, on the basis of close similarities to assemblages in south-central Alaska, Ackerman (1996c:462) assigned them to the Denali complex.

A similar assemblage was found in better stratigraphic context at Ilnuk on the Holitna River. A total of 4,750 artifacts, including more than 2,000 microblades, were excavated from loess deposits—averaging 30 centimeters in thickness—overlying a weathered limestone ridge. The occupation level lay beneath a volcanic ash horizon dating to ca. 4.4–4 ka (Aniakchak tephra) but failed to produce any older dates (Ackerman 1996d:466–467). The assemblage is classified as Denali complex and may date to the Younger Dryas or later (early Holocene?).

Cave 1, Lime Hills is located on the northern slope of the Alaska Range east of Lime Village and the Stony River at an elevation of 527 meters above sea level (Ackerman 1996a:470). Described in chapter 4, the small cave yielded possible traces of occupation as early as 18.9–16.1 cal ka in the form of modified caribou bones. The principal significance of Cave 1, Lime Hills, however, is the overlying component, which provides evidence for occupation at the end of the Younger Dryas or immediately thereafter. This component contains fragments of a microblade and bone point at depths of 36–48 centimeters in a sequence of cave deposits up to 1.2 meters in thickness (Ackerman 1996a:470–473). The microblade was associated with charcoal dated to 10.9 cal ka (i.e., following the end of the Younger Dryas). The bone (or antler) point fragment measures 10.7 centimeters in length and exhibits opposing lateral slots that were presumably used for microblade inserts (Ackerman 1996a:472, fig. 10-7). It is one of the few—and probably the oldest—examples of a slotted bone/antler point found in Alaska.

Northern and Central Alaska Range

The Alaska Range contains major sites dating to the Younger Dryas interval that are found both in the northern foothills zone and in the central part of the range. The most important groups of sites are concentrated in the Nenana Valley and in the Tangle Lakes region. As in the case of the Tanana Valley, many of these sites—especially those at high elevation—seem to have been occupied toward the end of the Younger Dryas. But unlike the Tanana Valley, where at least some of the sites may have been occupied during winter months, both the northern foothills and alpine areas of the Alaska Range probably were visited only during the summer and fall. As in historic times, the Alaska Range sites are likely to reflect a different aspect of the economy than those of the lowlands.

Field research in the Alaska Range was critical to development of interpretive and temporal frameworks for Beringian archaeology. Frederick Hadleigh West (1967) based his definition of the Denali complex primarily on sites in the Upper Teklanika Valley and the Donnelly Ridge site (Delta River).[10] As originally defined, the Denali complex comprises small wedge-shaped microblade cores, core tablets, microblades, biconvex bifacial knives, burins, end-scrapers, and other diagnostic items (West 1967:371–372). The burins often exhibit multiple spall removals ("Donnelly burins") and resemble the microblade cores (West 1981:124–125). The Dry Creek site in the Nenana Valley provided critical support for the stratigraphic and temporal placement of this industry at the end of the Pleistocene (Powers and Hamilton 1978). West (1981:91–137) later excavated more Denali complex sites in the Tangle Lakes region of the south-central Alaska Range.

If the Brooks Range contains the highest concentration of Mesa complex sites—along with isolated fluted points—the "heartland" of the Denali complex may lie in the Alaska Range (Mason, Bowers, and Hopkins 2001:533). Sites of Younger Dryas age containing typical Denali artifacts also are found in other parts of the southern and central interior and along the southeastern coast of Alaska. As in the Tanana Basin, however, there are traces of the Mesa complex in the Alaska Range—most importantly at Dry Creek—and it is apparent that both industries are present during this interval.

NORTH ALASKA RANGE: DRY CREEK

The Dry Creek site contains a cultural layer (Component II) dating to the Younger Dryas interval that is as important to Beringian archaeology as the underlying occupation of Lateglacial age. Component II yielded a massive sample of stone artifacts (almost 29,000 artifacts), as well as identifiable large mammal remains, buried in deep stratigraphic

context that provided the first reliable early dates on the Denali complex (West 1975). In recent years, the character of the eastern Beringian
archaeological record during the Younger Dryas has become clearer,
and it is now apparent that this occupation layer is particularly representative, with evidence for bison hunting and coexistence of the Paleoarctic and Paleoindian traditions in the interior.

The Dry Creek site is located in the upper foothills on a glacial outwash terrace in the Nenana Valley (see chapter 5). Three occupation
levels are buried in almost 2 meters of loess and eolian sand that unconformably overlie the outwash gravels (Thorson and Hamilton 1977:156,
table 1). Component II is contained in a brown loess (Loess 3) that
consists of sandy silt with minor clay with discontinuous dark organic
lenses (Paleosol 1). The latter reflect immature tundra soil formation,
but Loess 3 is capped with a thicker and continuous organic horizon
(Paleosol 2) that represents a more strongly developed tundra soil
(Thorson and Hamilton 1977:161). Four radiocarbon dates obtained on
charcoal from Paleosols 1 and 2 yielded a range of 12.5–10.0 cal ka (Hoffecker, Powers, and Bigelow 1996:346, table 7-1).

Faunal remains were poorly preserved at Dry Creek and most were
unidentifiable, but a few large mammal teeth in Component II were
assignable to steppe bison (*Bison priscus*) and Dall sheep (*Ovis dalli*).[11]
The bison teeth included enamel fragments of premolars and molars
from a minimum of five individuals—three young and two adults
(Guthrie 1983a:242, table 6.2). The sample of sheep teeth from Component II was combined with the sample for Component I. In addition
to the mammal remains, avian gastroliths or gizzard stones were recovered during the final year of excavations, and several groups of them
were mapped on the Component II floor (Powers and Hoffecker
1989:277, fig. 7). As in the case of the gastroliths recovered from
Component I, they fall in the ptarmigan size range and probably were
deposited during summer, fall, or early winter (Guthrie 1983a:
274–281).

The Component II occupation floor was excavated over an area of
345 square meters and revealed fourteen concentrations of lithic debris
of varying size and composition (figure 6.10). The concentrations
range in area from approximately 4 to 13 square meters and included
between fewer than 350 to more than 3,500 artifacts. Most were associated with weathered mammal remains and traces of charcoal (Hoffecker,

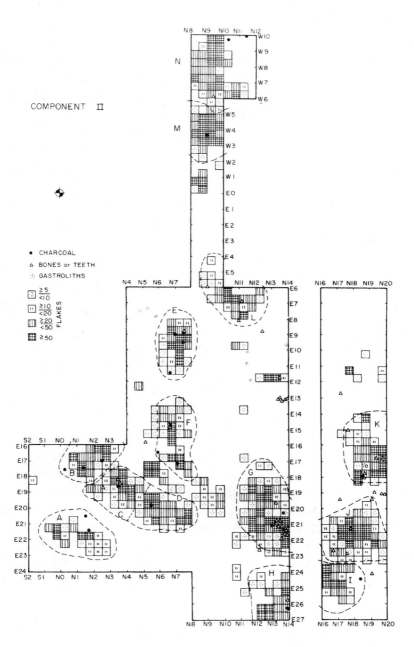

Figure 6.10 The distribution of artifacts, mammal bones and teeth, avian gastroliths, and charcoal fragments on the middle level at Dry Creek, which appears
to contain two distinct complexes (from Powers and Hoffecker 1989:277, fig. 7;
reproduced by permission of the Society for American Archaeology from *American Antiquity*, vol. 54, no. 2, 1989)

Powers, and Bigelow 1996:346–347). Five of the concentrations (A, B, C, G, and N) contained large numbers of microblades and microblade fragments. These concentrations also yielded microblade cores, core parts, and burins. Also present in Component II are biconvex knives, several of which are associated with two debris clusters (D and F). These artifact concentrations all contain diagnostic elements of the Denali complex (West 1967, 1981) and may be attributed to that industry. One of them (C or D) is associated with bison teeth, while several others (F, G, and N) are associated with unidentified mammal remains (figure 6.11).

Among the remaining debris concentrations, at least three (E, J, and K) contained bifacial projectile points, point tips, or point bases that are similar to point types found in the Mesa complex. They include a small concave-base lanceolate point (E), a thin, parallel-sided point tip (J), and several flat or expanding-stem point bases (J and K) (Powers 1983:125–131). These concentrations yielded small waste flakes produced by bifacial retouching, including delicate pressure-flaking. All of them are associated with mammal remains, which in one case (J) were identifiable as bison. Although originally assigned to the Denali complex along with the others (Powers and Hoffecker 1989:273–276), debris concentrations E, J, and K now are more logically attributed to the Mesa complex (figure 6.12).

The other debris concentrations on the Component II floor (H, I, L, and M) lack items diagnostic of either complex.[12] All of them did produce large and rather crude-looking bifacial tools or their fragments, and many of the waste flakes appear to have been derived from manufacture or resharpening of these tools. Three (H, L, and M) were associated with unidentifiable mammal remains. Although originally included—along with all the other remains in Component II—with the Denali complex assignment (Powers and Hoffecker 1989), concentrations H, I, L, and M now are better left unclassified with respect to complex or tradition. They could conceivably be linked to either the Denali or the Mesa complex (or possibly neither).

Dry Creek Component II reveals a complicated picture of Younger Dryas occupation. People who manufactured typical Denali artifact types visited the site but apparently did not always make or use microblades here. People who made and used projectile points characteristic of the Mesa industry also were present at Dry Creek. Presumably, they account for at least some of the bison remains, which probably indicates

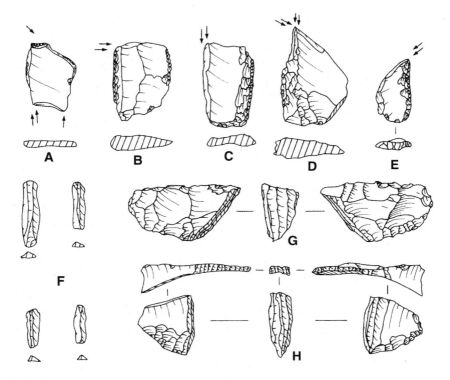

Figure 6.11 Burins, microblades, and microblade cores from the middle level (Component II) at Dry Creek assigned to the Denali complex (from Powers and Hoffecker 1989:274, fig. 5; reproduced by permission of the Society for American Archaeology from *American Antiquity*, vol. 54, no. 2, 1989)

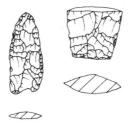

Figure 6.12 A small lanceolate point and point base from the middle level (Component II) at Dry Creek assigned to the Mesa complex (from Powers and Hoffecker 1989:275, fig. 6;reproduced by permission of the Society for American Archaeology from *American Antiquity*, vol. 54, no. 2, 1989)

that they came to the site in the fall or winter (Guthrie 1983a:244). One or both of these groups sometimes made and used heavy bifacial tools here—perhaps to butcher large mammal carcasses.

NORTH ALASKA RANGE: OTHER SITES

The Moose Creek site rests on the highest terrace level in the Nenana Valley—composed of Tertiary gravels and more than 200 meters above the modern floodplain—roughly 20 kilometers downstream from Dry Creek (see chapter 5). A series of occupation levels are buried in loess and eolian sand that varies 80–40 centimeters in thickness and unconformably overlies the ancient gravels. The second lowest level (Component II) lies in a unit of tan loess and is associated with faint discontinuous organic lenses that probably represent a weakly developed tundra soil similar to Paleosol 1 at Dry Creek (Thorson and Hamilton 1977:161). A charcoal sample from a partially excavated former hearth on this level yielded an age of 12.4 cal ka (Pearson 1999:334, fig. 2).

Component II was identified during the 1996 excavations at Moose Creek (45 square centimeters exposed) and includes a small concentration of microblades, a burin, small bifaces and biface fragments, including possible point tips and bases (Pearson 1999:336–338). At least some of the diagnostic elements of the Denali complex are present, and the assemblage is assigned to this industry (Pearson 1999:340). Also present in the lower levels at Moose Creek are several lanceolate point fragments—similar to those assigned to the Mesa complex from Dry Creek—but these were excavated from a test trench in 1979, and their relationship to the Component II assemblage is unclear (Hoffecker 1996:366, fig. 7-16).

During the Younger Dryas interval, the Panguingue Creek site—located a few kilometers downstream from Dry Creek on the same outwash terrace level—was occupied apparently for the first time. The site was discovered in 1976 and subjected to some testing in 1977 and 1985. Excavations were conducted in 1991, exposing a total of approximately 100 square meters (Goebel and Bigelow 1996: 366–367). Although up to 2 meters of eolian sand and loess overlie the outwash gravels, the three occupation levels are confined to the uppermost 50 centimeters of sediment. Unlike most localities in the northern foothills of the Alaska Range, there is a relatively thick deposit of late

Pleistocene sand and silt at Panguingue Creek that may have accumulated as early as 16 cal ka (Goebel and Bigelow 1996:368, fig. 7-17).

The lowest level (Component I) lies in a loess unit roughly 40–55 centimeters below the surface that contains two buried soil horizons, which appear to be analogs to the two lowest paleosols at Dry Creek (see above). The lowest buried soil comprises discontinuous organic lenses that are no more then 0.5 centimeters in thickness. The upper soil consists of a nearly continuous organic band between 1 and 10 centimeters in thickness (Goebel and Bigelow 1996:368). The older soil is undated, but the upper soil yielded three radiocarbon dates between 11.9 and 9.1 cal ka; two of the three dates fall into the later Younger Dryas. In contrast to Dry Creek, the occupation level in this time range is associated with the upper—not lower—buried soil and may therefore postdate Component II at Dry Creek. The radiocarbon dates also suggest a slightly younger age for this cultural level at Panguingue Creek.

No features or faunal remains were found in Component I, and the artifact assemblage was relatively small. In addition to a subprismatic core and approximately sixty waste flakes, the assemblage contained two lanceolate bifacial points of varying length, two transverse scrapers on flakes, and a boulder-chip scraper (Goebel and Bigelow 1996:369). The points exhibit rounded or subrounded bases. Although microblades are absent, the point types in Component I at Panguingue Creek conform to the types found at sites assigned to the Denali complex (West 1981:125). They are especially similar to bifacial points (or knives) recovered from artifact concentrations D and F at Dry Creek (Component II), which also are assigned to the Denali complex. There are no traces of the flat or concave-base point types found in the Mesa complex.

TANGLE LAKES REGION

The Tangle Lakes currently stand at about 850 meters above sea level in the central Alaska Range at the headwaters of the Delta River, which flows northward through the mountains and empties into the Tanana River. At the end of the Pleistocene, after local deglaciation—which occurred during the Lateglacial—lake waters were some 30 meters higher than they are today (West 1981:126–129). During Younger Dryas times (and perhaps for some period thereafter), the higher waters formed a large proglacial lake at the upper end of the basin, and sites occupied at the time lie above the former lake level (West 1996b:379–380).

Archaeological remains in the Tangle Lakes area were discovered in 1956 by William Hosley (West 1996b:378), and further investigations were undertaken by Ivar Skarland and Charles Keim (Skarland and Keim 1958). Survey and excavation were conducted by Frederick Hadleigh West, Douglas R. Reger, and colleagues during the 1960s (West 1996b:378). More than 100 sites are recorded, and although surficial sediment cover is thin, several have produced traces of occupation in dated stratigraphic context of Younger Dryas age. West (1981:135–137) used the results of his investigations in the Tangle Lakes to better define the Denali complex, and the sites in the area add significantly to an understanding of this industry.

The Phipps site (also known as Mt. Hayes III) is located on a sand feature linked to the former high lake level. The site was discovered and investigated in 1966–1968. A single cultural layer was defined at between 30 and 40 centimeters below the surface (West, Robinson, and Curran 1996:382, fig. 8-1). The artifacts were recovered from a sandy loess unit and associated with a buried soil horizon. Three radiocarbon dates indicate an age range similar to that of Component I at Panguingue Creek (12–9 cal ka [Mason, Bowers, and Hopkins 2001:528]), and dates from layers above and below the cultural level provide additional support for the chronology (West, Robinson, and Curran 1996:384, table 8-1).

The assemblage from Phipps comprises 22 wedge-shaped microblade cores, along with numerous core parts and more than 350 microblades. Also present are 23 burins and 81 burin spalls (West, Robinson, and Curran 1996:381–384).[13] Wedge-shaped microcores, core parts, microblades, and burins (some of which are multifaceted) are among the most characteristic forms of the Denali complex (West 1981:115–126), but they are typically found with bifacial tools and other items. The exclusive focus on microblades and burins at Phipps is unusual, although it may be seen in artifact concentrations A, B, C, G, and N in Component II at Dry Creek (see above). It presumably reflects production or repair of slotted bone or antler points with microblade insets (West, Robinson, and Curran 1996:384).

Located roughly 250 meters east and south of the Phipps site, Whitmore Ridge provided a somewhat different perspective on the Denali complex. At this location, the eolian deposits that contain the artifacts overlie an esker complex. No more than 30 centimeters of loess rests on fluvial sand and gravel of the latter. The cultural layer—subdivided into two components—is buried in the lowermost 10 centimeters of the loess and associated with a paleosol (West, Robinson, and West 1996:388, fig. 8-4). The lower component rests within a Bb horizon, and the upper component is associated with the overlying A2b horizon (West, Robinson, and West 1996:393). Five radiocarbon dates from the former suggest a Younger Dryas age (12.6–10.9 cal ka). The upper component presumably postdates the Younger Dryas.

Two concentrations of artifacts were found in the lower component. One of these concentrations ("Locus 1") was small and confined primarily to wedge-shaped microblade cores, core parts, and microblades (burins were absent). The other concentration ("Locus 3") comprised a mass of bifacial thinning flakes, large blades, and several tools—burin, scraper, and bifaces. Some microblades apparently are present as well. Among the bifaces are diagnostic Denali forms (e.g., West, Robinson, and West 1996:392, fig. 8-7c). Once again the pattern is similar to that observed at Dry Creek.

The upper component contains one artifact concentration ("Locus 2") that includes blades and microblades, along with small conical blade cores (West, Robinson, and West 1996:389, fig. 8-5a–c), biface fragments, and one burin. This assemblage is similar to that of the more or less contemporaneous Component II at Panguingue Creek (Goebel and Bigelow 1996:369–370). They apparently represent a post–Younger Dryas industry of interior Alaska that may be related to the older Denali complex but also bears similarities to the Sumnagin culture of Northeast Asia (West, Robinson, and West 1996: 393–394).

Pacific Coast: Southeast Alaska

One of the most dramatic changes in the archaeological record of the Younger Dryas is the appearance of sites in the coastal zone. The earliest known sites are found in southeast Alaska (e.g., Ground Hog Bay 2

[Ackerman 1996e]). Although early Holocene settlement in the Aleutians was documented many years ago (Laughlin 1975), older sites have yet to be found in this region or on the southern coast of Chukotka and Kamchatka. In southeast Alaska, however, several occupations have been dated to the end of the Younger Dryas or its immediate aftermath—as early as 11.7–11.3 cal ka.

Many archaeologists have long suspected that older sites were occupied along the former southern coast of Beringia but were then inundated by rising sea levels at the end of the Pleistocene. New research on the glacial chronology of the Northwest coast indicates that deglaciation of some areas occurred relatively early—opening coastal habitats to settlement by 17–16 cal ka (Mandryk et al. 2001: 304–307). In fact, early settlement of the south Beringian and Northwest coasts may be the only plausible explanation for the presence of sites south of the Canadian ice sheets before ca. 15–14 cal ka (i.e., before the existence of an ice-free corridor in the interior [Dixon 2001; Dyke 2004]).

The early sites in southeast Alaska contain wedge-shaped microblade cores and microblades, and the assemblages have been compared to those of the Denali complex and other Paleoarctic tradition industries of the interior (e.g., Ackerman 1996e:429). Although some believe that an older industry without microblade technology was established at an earlier time on the coast (e.g., Carlson 1990), this remains to be firmly documented. The similarities between the coast and interior—and the greater antiquity of the interior microblade technology—have encouraged speculation that the interior technology is the source of the coastal industry (e.g., West 1996a:549). Noting significant differences between the early Holocene microblade assemblages of southern Alaska and those of the interior, Steffian, Eufemio, and Saltonstall (2002) have disputed this suggestion.

Perhaps the most important find in southeast Alaska has been the recovery of human skeletal remains of early Holocene age from On-Your-Knees Cave (49-PET-408) on Prince of Wales Island. Stable isotope analysis of the bone indicated a diet based primarily on marine foods (Dixon 1999:117–118). This confirmed the assumption that—despite its close similarity to the Denali complex—the coastal microblade technology had indeed been adapted for use in a maritime economy.

Conclusions

The Beringian archaeological record of the Younger Dryas differs significantly from that of the final Lateglacial. In place of the uniquely Beringian but somewhat obscure Lateglacial industry, the Younger Dryas sites contain traces of two sharply defined traditions (or complexes) that are clearly derived from areas outside Beringia. The contrast between the Lateglacial and Younger Dryas could be influenced by biases created by the comparatively small sample of sites assigned to the former. A larger sample of Lateglacial assemblages[14] may eventually reveal greater similarity with those of the Younger Dryas. This seems especially likely with respect to similarities between the Lateglacial industry and the microblade technology or Paleoarctic Tradition of Younger Dryas times. At present, however, the contrast appears to be considerable.

What accounts for this apparent contrast in the archaeological records of the Lateglacial and Younger Dryas periods? At least part of the difference is almost certainly a consequence of colder and drier climates during the latter. Younger Dryas climates expanded steppic habitat and promoted a significant increase in the steppe bison population. This apparently accounts for the presence of technologies in Beringia—and presumably the people who made them—related to bison hunting on the Great Plains. The late Paleoindian complex of Beringia represents a striking reversal of the normal pattern of movement into higher latitudes during intervals of warmth (Hoffecker 2005a).

Bison hunters brought diagnostic Plains point types such as Agate Basin and Hell Gap—as well as some fluted forms, it seems—northward along the eastern slope of the Canadian Rockies as far as the northern Yukon and Brooks Range in northern Alaska (e.g., Dumond 2001). They also penetrated into the Tanana Valley and other parts of central Alaska but did not move south of the Alaska Range into the coastal zone. Their absence in western Beringia[15] probably reflects inundation of the land connection between Chukotka and Alaska. As climates warmed—and bison numbers eventually declined—after the Younger Dryas, the Paleoindian assemblages vanished from the northern archaeological record (Mann et al. 2001:133).

Accounting for the apparent rise in Paleoarctic assemblages containing microblades is more difficult. In contrast to the Paleoindian occupations, these assemblages are more plausibly related to the preceding Lateglacial industry of Beringia, which contains at least some microblade technology (described in chapters 4 and 5). Perhaps the Paleoarctic industry of the Younger Dryas also reflects changes related to climate. Along these lines, Mason, Bowers, and Hopkins (2001:539) suggested that the increase in microblade assemblages might have been tied to greater availability of caribou—due to colder conditions—but supporting faunal evidence from the sites is limited.[16] An alternative explanation is that, like the Paleoindian complex, the Paleoarctic industry was imported by new arrivals. The sunken entrance tunnels at Ushki seem to indicate more direct technological responses to lower winter temperatures (Dikov 1977; Hoffecker, Powers, and Goebel 1993:53).

Although Northeast Asia is the most obvious source for a new migration of microblade-making people into Beringia during the Younger Dryas, regional shifts within Beringia also might account for changes in the archaeological record at this time. William S. Laughlin (1967:447) emphasized the flooding of the land bridge as a stimulus to the movement of people, and current sea-level chronology suggests that this was a Younger Dryas event. The Paleoarctic industry may have been established at some earlier point along both the south coast of the land bridge (Laughlin 1967:423–427) and other southern continental shelf areas that were inundated during Younger Dryas times. The industry is documented in the coastal zone by 11.7 cal ka. Unlike the Paleoindian complex of northern and central Alaska, the microblade industry carried on into the post–Younger Dryas era (West 1996a:551; Goebel and Slobodin 1999:146–147).

Beringia and the New World

Despite the remoteness and lack of development, much field research has been conducted on the earliest prehistory of the surviving land portions of Beringia. Since the 1930s, archaeologists on both sides of the Bering Strait have been searching for remains of the first Beringians. And much of the impetus for this search—including archaeologists in Northeast Asia—has derived from an almost obsessive concern with the origins of the native population of the Americas.

As described in chapter 1, speculation on the origins of Native Americans has focused on Northeast Asia and Alaska since Fray José de Acosta published *Historia natural y moral de las Indias* in AD 1590 (Wilmsen 1965). During the late nineteenth and early twentieth centuries, archaeologists debated the question of the antiquity of New World peoples. The discovery of fluted projectile points in association with extinct bison and mammoth—and the subsequent application of radiocarbon dating to sites containing such points—eventually revealed that Paleoindians had been present at least as early as the end of the Pleistocene (Fagan 2004). As the concept of Beringia and the land bridge emerged in the late 1930s, it was incorporated into thinking about the timing and routes of the initial colonization of the Western Hemisphere (Hopkins 1967a:3–4).

In addition to the presence of the land bridge during intervals of lowered sea level, two other concepts were important to Beringia's perceived role in the peopling of the New World. The first of these was the idea of a massive steppe-tundra biome extending from northern Eurasia across the land bridge into Alaska during glacial periods (e.g., Guthrie 1982:308, fig. 1). The second was the "ice-free corridor" between the Laurentide and Cordilleran ice sheets (e.g., Mandryk et al. 2001). According to what is sometimes characterized as the "classic model," Native Americans—excluding the more recent peoples of the arctic and subarctic regions—are derived from mobile bands of big-game hunters who colonized the Beringian "mammoth steppe" at the end of the Pleistocene and moved into similar habitat on the North American Plains as retreating ice sheets opened interior access to the latter from eastern Beringia (e.g., West 1981:188–206; Haynes 1982:396–398).

In recent years, many—perhaps most—archaeologists have rejected the classic model because widely accepted dates on one or more sites located south of the ice sheets are older than the current evidence for the opening of the corridor route between the latter (e.g., Dixon 2001:277–278). From an archaeological perspective, however, the shortcomings of the Beringian interior model are more long-standing and fundamental. Archaeologists never have succeeded in documenting the presence of artifacts in Beringia that are very similar—of comparable or greater age and thus likely related closely—to the oldest firmly dated Paleoindian artifacts found south of the former ice sheets (Dumond 1980). It is for this reason that the debate over the origins of Native Americans—and the relevance of Beringia to that problem—continues to rage.

As early as 1935, Nels Nelson reported the discovery of microblade cores from central Alaska (Campus site) that established a link between the early occupants of eastern Beringia and the Upper Paleolithic of northern Asia (Nelson 1935:356). Radiocarbon dates from the lowest level at Swan Point in the Tanana Valley suggest that Nelson had indeed found artifacts characteristic of the earliest people in Alaska (Holmes 2001; C. E. Holmes, pers. comm., 2006). But while the artifacts are indistinguishable from those of the Lena Basin at 15 cal ka, they are unknown south of the ice sheets. The oldest eastern Beringian industry is tied closely to the late Upper Paleolithic of Siberia—not to the early Paleoindian complex of the North American Plains. Although Frederick Hadleigh West (1981, 1996a) has long argued that the latter may be derived from the former, most archaeologists believe that the differences between the two industries are considerable.

A few years after Nelson described the microblade cores from central Alaska, Froelich Rainey (1939) reported several lanceolate projectile points from the Fairbanks muck. These points, which resemble some nonfluted Paleoindian point types of the Plains, were found in association with late Pleistocene large mammals—although the relationship between the artifacts and bones was problematic (Rainey 1940:305–307). Even more dramatic finds occurred during 1947–1955, as various investigators recovered fluted points and point fragments from shallow and surface contexts in northern Alaska and the Yukon (Thompson 1948; Solecki 1951; MacNeish 1956). The design of these artifacts suggested possible links with the oldest known Paleoindian complexes of the Plains (i.e., Clovis and Folsom). But after many years of further discovery and research, it has become clear that most of the fluted and nonfluted lanceolate points in eastern Beringia postdate 12,000 calibrated years ago (12 cal ka). They apparently represent—as proposed many years ago—a relatively late northward movement of Plains technology that was probably tied to the Younger Dryas cold interval and the expansion of bison habitat in northwest Canada and parts of Alaska (Mann et al. 2001:132).

In more recent decades, several researchers have suggested that Beringia contains a unique and perhaps indigenous industry that may represent the source of the earliest cultures in North America. The late N. N. Dikov (1979:31–53) proposed that the oldest assemblage at Ushki, which comprises stemmed bifacial points, burins, and end-scrapers, was the progenitor—via an interior route—of the earliest industry at Marmes Rockshelter (Washington) and other early sites in northwestern North America. Both the lowest level at Ushki and sites containing similar stemmed points in the Northwest subsequently were re-dated to younger time periods (Goebel, Waters, and Dikova 2003). Nevertheless, the artifacts in Layer VII at Ushki remain an important part of the early Beringian record.[1]

Along similar lines, Hoffecker, Powers, and Goebel (1993) argued that several assemblages in central Alaska containing small bifacial

points, end-scrapers, and other items (Nenana complex) might represent a source for the early Paleoindian complexes of the Plains (see also
Goebel, Powers, and Bigelow 1991). In the early 1990s, it appeared that
the Nenana complex might date to as much as 14 cal ka and represent
the oldest Beringian industry (Powers and Hoffecker 1989). The later
discovery of an earlier microblade assemblage in the Tanana Valley and
more precise dating of the Nenana occupations to ca. 13 cal ka (Hamilton
and Goebel 1999:164) convinced many researchers that the latter were
probably unrelated to the early Paleoindian sites south of the ice sheets
and perhaps merely a functional subset of the Beringian microblade
industry (e.g., Meltzer 2001:211–212; Adovasio and Pedler 2004).

The purpose of this final chapter is to assess the contribution of the
Beringian record to the problem of the initial colonization of the New
World. As currently known and interpreted, how do the Beringian data
help to evaluate the various models of New World settlement? The
contents of the preceding chapters—the human ecology of Beringia—
are summarized. Then, each of the principal models of New World
settlement is discussed in terms of the Beringian sites and their context.

Human Ecology of Beringia: A Summary

Among the hominins, only modern humans (*Homo sapiens*) could
adapt to environments above latitude 60°N—thus excluding earlier
humans from Beringia. Although the Neandertals evolved a variety of
anatomical and foraging adaptations to relatively cold and unproductive environments in northern Eurasia, they were unable to cope with
most Northeast Asian habitats due to extreme winter temperatures or
low density of large mammals, or perhaps a combination of the two
(Goebel 1999:212–213; Hoffecker 2005a:53–69). They also seem to have
abandoned the colder and drier areas of Europe during glacial advances
(Gamble 1986; Hoffecker 2002a).

Modern humans present a stark contrast to the Neandertal pattern,
and the human ecology of Beringia begins in Africa and not in Siberia.
Dispersing out of Africa after 60 cal ka, modern humans rapidly
occupied the East European Plain and southern Siberia, despite anatomical adaptations to warm climates (Hoffecker 2005a). Their ability
to design complex technology—including tailored clothing, heated

shelters, and novel hunting and trapping devices—probably was the critical factor, although organizational strategy also may have been important. In the broader context of their colonization of northern Eurasia, modern humans apparently became the first hominins to expand above 60°N and beyond. At the end of the interstadial period that is the age equivalent of MIS 3 (ca. 60–28 cal ka), their sites are found on and above the Arctic Circle (66°N) in Europe (Pavlov, Svendsen, and Indrelid 2001), and as far as 71°N in Northeast Asia (Pitul'ko et al. 2004).

In the lower Yana River Basin, modern humans occupied a site (Yana RHS) that lies within the boundaries of Beringia, as defined in chapter 1. Unfortunately, very little is known about the people at Yana RHS and other sites of comparable age (40–30 cal ka) above latitude 60°N. The sites contain few artifacts and no traces of former shelters, and it is not clear whether they represent seasonal movements into high latitudes or year-round settlement of the Arctic. Both large and small mammal remains were found associated with the artifacts at Yana RHS (Pitul'ko et al. 2004:55). Sites like these undoubtedly suffer from low archaeological visibility because of their age, size, and remote locations, and presumably many more of them remain to be discovered. During the final phase of MIS 3—when Yana RHS was occupied—climates were relatively mild in northwestern Beringia, and both woody shrubs (birch) and trees (larch) were growing in lowland areas (Anderson and Lozhkin 2001:98–99).

Climates became significantly colder in the Northern Hemisphere during the early phase of the MIS 2 age equivalent (roughly 27–23 cal ka). There is no evidence for sites—even small seasonal encampments—above 60°N after 27 cal ka, but sites in mid-latitude regions of northern Eurasia (45–55°N) are often large and suggest increased group size and extended occupations, if not higher population density (Svoboda, Lozek, and Vlcek 1996:131–170; Goebel 1999:216–218). A new wave of technological innovations may have contributed to the pattern by facilitating further expansion of dietary breadth (e.g., waterfowl) and greater foraging efficiency (Hoffecker 2005b).

Despite the increased numbers and dimensions of sites during the early phases of MIS 2, as the glaciers in the Northern Hemisphere reached their maximum extent (the Last Glacial Maximum, or LGM) areas of northern and eastern Europe, as well as much of Siberia, may have been largely or wholly abandoned for more than a thousand years (23–21 cal ka) (Tseitlin 1979:260; Goebel 1999:218; Hoffecker 2002a:200–201). There is no evidence for occupation of Beringia during this interval. The reason(s) for settlement decline might lie in reduced biotic productivity (due primarily to lower moisture), a shortage of available fuel sources (including possibly woody shrubs), or extreme low winter temperatures, or all three. As we have suggested elsewhere (Hoffecker and Elias 2003:37–39), modern humans living in northern Eurasia before 20 cal ka probably were susceptible to extreme cold because of their warm-climate anatomy, and this may have been an important constraint on high-latitude settlement.

The reoccupation of Northeast Asia took place as climates began to warm after 20 cal ka, and the size and distribution of sites reveals a pattern that is very different from that of the pre-LGM period. Sites in southern Siberia are relatively small and probably represent short-term occupations. The pattern has been characterized as a "microblade adaptation" (Goebel 2002) and apparently reflects high mobility and small group size. The wedge-shaped microblade cores so characteristic of these sites suggest an appropriately portable and parsimonious technology. The faunal remains indicate consumption of various large mammals such as reindeer, bison, sheep, and red deer (elk), as well as some small mammals and birds (Goebel 1999:220–221). The contrast with Europe—which contains many large sites and extended occupations during 20–15 cal ka—is striking and presumably is a function of lower productivity in Siberia (Hoffecker 2005a:110–114).

The occupation—or reoccupation—of Beringia is dated to 15–14 cal ka and coincides with a transition from arid steppic environments that supported a diverse array of large mammal grazers ("mammoth fauna" [Guthrie 1982]) to mesic shrub tundra. Although the Lateglacial settlement of Beringia may be placed into the broader context of the reoccupation of Northeast Asia after the LGM, it appears likely that local environmental variable(s)—related to the shrub tundra transition—opened areas east of the Lena Basin for settlement at this time. The most important factor may have been the significant increase of woody shrubs in the form of willow and dwarf birch, which provided more available fuel, including "starter fuel" for fresh bone: traces of burned bone are common in Beringian sites (e.g., Cook 1969; Mochanov 1977; Goebel and Slobodin 1999; Holmes and Crass 2003).

Although the sample of sites is extremely small, it presently appears that the Lateglacial settlement of Beringia follows the spread of the shrub tundra from the lowlands to higher elevations (Bigelow and Powers 2001). The earliest sites are found in the Indigirka lowland in northwestern Beringia (Berelekh) and the Tanana Basin in central Alaska (Broken Mammoth, Swan Point, and Mead) dating to 14–13.5 cal ka (Mochanov and Fedoseeva 1996c; Holmes 2001). Younger Lateglacial sites dating to about 13 cal ka materialize in the Upper Tanana and northern foothills of the Alaska Range (e.g., Dry Creek and Walker Road) as the shrub tundra spread to upland zones. The Bluefish Caves at 600 meters above sea level in the northern Yukon also may have been occupied at this time.

Faunal remains in the Lateglacial Beringian sites confirm that their occupants were pursuing a shrub tundra economy based on large mammals of the postglacial era such as elk or wapiti (*Cervus* sp.) and sheep (*Ovis dalli*), along with various small mammals (especially hare), numerous birds, and some fish (Hamilton and Goebel 1999:184; Yesner 2001). Small mammals may have been a significant resource, as during late Holocene times in the same regions (Stepanova, Gurvich, and Khramova 1964; Nelson 1973).[2] Waterfowl seem to have had an especially important role in the diet—reflecting expansion of freshwater aquatic habitat in a rapidly thawing Beringian landscape (Bigelow and Edwards 2001:211; Mann et al. 2001:125). Although some have emphasized the role of the "mammoth fauna" in the economy (e.g., West 1996a:551), it is apparent that most of these taxa became rare or extinct during the Lateglacial (Guthrie 2003, 2004, 2006). Mammoth remains are present at some sites but may represent bones and tusks scavenged from natural occurrences.

In many respects, the Lateglacial Beringian economy had already acquired a Holocene character—similar to that of northern Athapaskan peoples of the Alaskan/Yukon interior, but with perhaps less reliance on fishing (e.g., Nelson 1973). The economy also was similar to that of recent peoples of interior Northeast Asia such as the Yukaghir, but apparently with less dependence on reindeer hunting (Forde 1934:101–106; Stepanova, Gurvich, and Khramova 1964). Population density must have been low—as among the northern interior peoples mentioned above—and large aggregations of people probably rarely occurred. Sites presumably were occupied by small groups on a temporary basis to exploit seasonally available resources such as water-fowl in the spring (e.g., Healy Lake) and elk in the fall (e.g., Dry Creek). The scarcity of wood—indicated by heavy use of bone and possibly other alternative fuels—may have placed some special constraints on settlement location and scheduling.

The character of the material culture during the Lateglacial remains a major issue in Beringian archaeology. The earliest firmly dated sites in central Alaska contain a wedge-shaped microblade core and micro-blades very similar to those found in the Siberian interior (e.g., in the Dyuktai culture [Mochanov 1969; Powers 1973]), and this industry is clearly present during Younger Dryas times (after 12.8 cal ka). But assemblages dating to between 13.5 and 12.8 cal ka yield a more eclectic array of stone tools, and microblades are not particularly common. Especially diagnostic for this interval are small bifacial points of varying form ("Chindadn points") first recognized at Healy Lake that may have been used to tip throwing darts (Cook 1969:184–187). Stemmed points are found at Ushki on Kamchatka (Dikov 1979). End-scrapers are com-mon in several assemblages (e.g., Walker Road [Goebel et al. 1996]). A few microblades are reported in association with Chindadn points at Healy Lake (Cook 1996), and microblade cores and microblades are present at Bluefish Caves, which may have been occupied at this time.

Most of the technology of the Lateglacial Beringians has not been preserved. Ethnographic data indicate that technological complexity among hunter-gatherers is an inverse function of temperature and that the most complex technology is found at high latitudes (Oswalt 1976). Much of this technology is composed of bone, antler, ivory, hide, and various plant materials (Osgood 1940) that are rarely preserved in archaeological sites (because they are often affected by acidic boreal for-est soil formation of the middle and later Holocene). Beringian cloth-ing and shelter technology was probably especially sophisticated, given the demands of the subarctic climate (Oswalt 1987), but the economic emphasis on small mammals, birds, and fish also indicates a complex food-getting technology comprising nets, snares, throwing darts, fishing gear, and other instruments (Oswalt 1976). Transportation technology may have included sleds and other devices.[3]

Several archaeologists have proposed that at least some of the Lateglacial assemblages represent a unique complex that may be dif-ferentiated from the Dyuktai culture and other microblade industries (e.g., Dikov 1979; Powers and Hoffecker 1989). West (1996a:549–552) places all of these assemblages—along with those of the early Holocene period—in a pan-Beringian industry (i.e., Beringian tradition), whereas others recognize some significant changes in assemblage composition between the Lateglacial and the Younger Dryas interval but lump all of the former into one entity (e.g., Holmes 2001:156). However the final Lateglacial sites are classified—as one complex or more—they appear to represent a unique Beringian "archaeological culture" that undoubtedly reflects in part the peculiar demands of high-latitude shrub tundra environments. The character of the younger Lateglacial Beringian industry may also reflect isolation from adjoining regions: this industry does not bear a close resemblance to either Northeast Asian or mid-latitude North American assemblages of the period.

The archaeological record of Beringia exhibits significant changes with the onset of colder climate during the Younger Dryas (12.8–11.3 cal ka). Inundation of the land bridge apparently occurred during this interval, and a divergence between west and east is apparent for the first time. The most striking change in eastern Beringia (Alaska and Yukon) is the sudden appearance of assemblages containing lanceolate projectile points similar to those of the later Paleoindian complexes of the North American Plains. The point types include Angostura, Agate Basin, and Plainview, as well as some fluted forms (often exhibiting multiple flutes). These assemblages seem to be associated with hunting steppe bison (*Bison priscus*), which experienced a substantial increase in numbers during the Younger Dryas (Guthrie 1990), presumably due to the expansion of steppic habitat at the expense of shrub tundra (Bigelow and Edwards 2001:211–212).

The assemblages containing lanceolate points—many of which have recently been assigned to the Mesa complex (Kunz and Reanier 1994)—are especially common in the Brooks Range and other parts of northeastern Beringia (Alexander 1987; Reanier 1996).[4] Some of them are also found, however, in southwestern and central Alaska (e.g., Spein Mountain [Ackerman 2001]), including the northern foothills of the Alaska Range (e.g., Dry Creek, Component II). Many of the sites occupy prominent bedrock knobs that offer broad vistas of the surrounding landscape.

Although most of the Mesa assemblages lack faunal remains, some localities yielded bison teeth and bones (e.g., Engigstciak

[Cinq-Mars et al. 1991]), and this—in conjunction with the association between the Plains point types and bison procurement on the Plains —suggests the link with bison hunting (Mann et al. 2001:132–133). Kunz, Bever, and Adkins (2003:59–60) recently suggested that the Mesa complex was a local Beringian development, but the earlier history of lanceolate point technology and bison hunting on the Plains supports the idea that these assemblages were produced by people who had moved northward through the ice-free corridor (Dumond 2001). They may have penetrated into central and southwestern Alaska down the Porcupine River to the Yukon-Tanana Basin (figure 7.1).

At the same time, assemblages containing wedge-shaped microblade cores and microblades became common on both sides of the flooded Bering Strait, although they seem to be scarce above the Arctic Circle. In Northeast Asia, these assemblages are assigned to the Dyuktai culture (e.g., Ushki, Layer VI [Dikov 1977]), but in Alaska many of them are classified as Denali complex (West 1967). The microblades were inserted into lateral slots in bone points, and isolated examples of the latter have been found at several sites (e.g., Cave 1, Lime Hills [Ackerman 1996a:472, fig. 10-7]). In addition to the microblades, Denali assemblages often contain multifaceted burins and small bifaces (typically biconvex in cross-section) (West 1981:115–126). Denali complex sites occupy a variety of topographic settings in lowland basins (e.g., Swan Point, Cultural Zone 3 [Holmes 2001]), as well as upland zones (e.g., Moose Creek [Pearson 1999]), and they sometimes co-occur with the Mesa complex assemblages—most notably at Dry Creek.

If there are uncertainties about the full scope of the Mesa economy,[5] it appears that the people who produced the Denali/Dyuktai complex were pursuing an economy similar to that of their predecessors during the Lateglacial. At Ushki, Layer VI there is evidence for consumption of large mammals, small mammals, birds, and fish (Dikov 1979; Goebel, Waters, and Dikova 2003:503), and the Alaskan sites reveal a similar picture (Mason, Bowers, and Hopkins 2001:531; Yesner 2001). Many of the sites in upland areas may have been used primarily for large mammal hunting.

During the Younger Dryas period, traces of occupations along the southern coast of Beringia and adjoining areas materialize for the first time. The lowest component at Ground Hog Bay Site 2 on the Chilkat Peninsula in southeast Alaska dates to 11.5–10 cal ka and contains wedge-shaped microblade cores similar to those of the Denali complex

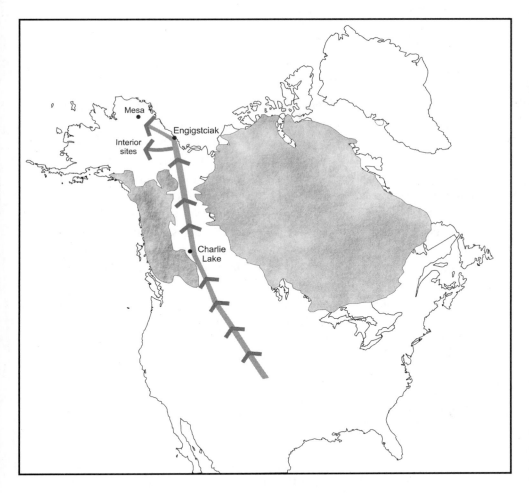

Figure 7.1 Northward movement of Paleoindian bison hunters from the plains through the ice-free corridor to eastern Beringia during the Younger Dryas

(Ackerman 1996e:426–429). Stable isotope analysis of human remains from a slightly younger occupation at On-Your-Knees Cave (Prince of Wales Island) indicates a heavy marine diet (Dixon 2001:286–287). This suggests that the south Beringian coast was occupied at an earlier date and that—as long noted by archaeologists—older sites probably lie in inundated continental shelf areas.

Some new technologies seem to have been developed during the Younger Dryas, and at least some of them appear to be a response to cooler climates. In contrast to the Lateglacial occupation, the former houses at Ushki Layer VI exhibit sunken entrance tunnels that would have reduced cold air flow into the living chamber (Dikov 1977:53, fig. 11). Ushki also yielded some large spatula objects of bone that may have been snow knives, while an eyed needle of bone was recovered from Broken Mammoth (Dikov 1979:68, fig. 24; Holmes 1996:316, fig. 6-6q). The first traces of domesticated dog in Beringia also are dated to the Younger Dryas—again at Ushki, Layer VI (Dikov 1979:60, fig. 15).

During the Younger Dryas period—probably earlier rather than later in the course of this interval—Beringia ceased to exit as a contiguous land mass. Climates warmed significantly after the end of the Younger Dryas (ca. 11.3 cal ka), and more changes are apparent in the archaeological record. In areas that had been part of western Beringia, Dyuktai assemblages were succeeded by the Sumnagin culture, also based on the production of microblades and reflecting a northern interior economy (Hoffecker 2005b:112–114). The steppe bison population in the former eastern Beringia probably declined at this time, and the Mesa assemblages seem to disappear altogether (Mann et al. 2001). Assemblages containing microblades (many of which are assigned to the Denali complex) are found in many coastal and interior regions, but there is some evidence of decline in the aftermath of the Younger Dryas, perhaps due to less favorable conditions for caribou (Mason, Bowers, and Hopkins 2001:540–542).

Models of New World Settlement: The Beringian Evidence

There are three models of New World settlement presently under consideration, and each of these is described briefly below. Although the fundamental strengths and weaknesses of each model are noted, the objective of the discussion is not to evaluate the models but to consider each in light of the current data from Beringia. None of the three models can be firmly refuted at present, and more than one of them might be valid.

TRANSATLANTIC OCEAN MODEL

Although a majority of North American archaeologists seem to view the possibility of transatlantic colonization of the Western Hemisphere during the Pleistocene as remote (e.g., Straus, Meltzer, and Goebel

2005),[6] this model has received increased attention in recent years and warrants consideration. The idea has been developed and advocated by Dennis Stanford and Bruce Bradley (Hadingham 2004; Bradley and Stanford 2004). The basic premise is that the earliest occupants of the New World were derived from people who manufactured artifacts of the Solutrean culture of southwestern Europe and who navigated the North Atlantic Ocean during the LGM (figure 7.2).

The chief advantage of this model is that it identifies a cultural source for the earliest well-defined Paleoindian complex in mid-latitude North America. It avoids the problem of deriving the Clovis complex from highly dissimilar industries in Northeast Asia and Beringia. Stanford

Figure 7.2 The Transatlantic Ocean model of New World settlement

and Bradley argue that Clovis lithic technology shares diagnostic attributes with the Solutrean, especially related to the production of parallel flaking of laurel-leaf points found in both cultures. Both Solutrean and Clovis point makers employed an unusual technique (controlling overshot flaking) for detaching flakes from one side by striking the opposite side of the point (Hadingham 2004:96).[7] Another advantage of this model is that it could account for the morphology of some early human skeletal remains in the New World that exhibit major differences from those of living Native Americans (e.g., Chatters 2000).

The disadvantages of the model are the significant temporal gap between the Solutrean (ca. 25–20 cal ka) and the Clovis complex (13 cal ka), and the undeniable difficulties of a transatlantic voyage during the Last Glacial Maximum with the apparently limited seafaring technology of the European Upper Paleolithic. Although advocates of the Solutrean-Clovis link have noted that several sites in the southeastern United States (e.g., Cactus Hill [Virginia]) may indicate the presence of people in North America during 20–13 cal ka, there is no sequence of firmly dated assemblages that documents the development of laurel-leaf points in this time range. Perhaps an even greater concern is the lack of evidence—either direct or indirect—for maritime adaptations in the Solutrean culture (Straus 2000:222–223).[8] There is no reason to believe that the Solutreans were constructing any form of watercraft, let alone one that would have facilitated a voyage across 5,000 kilometers of ocean and ice.

How do the Beringian data contribute to assessing the transatlantic model? In an odd way, the Beringian record has been critical to the genesis of this proposal. It is the lack of any clear link between the earliest well-defined (and well-dated) Paleoindian complex in mid-latitude North America and the sites in Beringia and Northeast Asia—coastal or interior—that seems to have been the primary catalyst for the transatlantic model (Hadingham 2004:94). This model reflects frustration with the absence of such a link, despite many decades of research, and it is a measure of the distance to which archaeologists may be willing to go in order to explain it.

PACIFIC COAST MIGRATION MODEL

The idea that most Native Americans might be descended from a population that migrated along the southern rim of Beringia and down a partially deglaciated Northwest coast has been under discussion for many years. The coastal migration model was described by Knut Fladmark (1978, 1979, 1983) and subsequently advocated by E. James Dixon (1999, 2001) and others (e.g., Gruhn 1994). Support for the model has grown in recent years as a consequence of several developments. Radiocarbon dates for occupation at Monte Verde in southern Chile suggest that people lived south of the Canadian ice sheet before a viable interior corridor route (Dixon 2001:277–279). At the same time, new studies of sea-level history and glacial chronology in the Northwest indicate that deglaciated coastal margins and food resources were available earlier than believed in the past (Mandryk et al. 2001:304–307) (figure 7.3).

As envisioned by Fladmark (1978:123–126), migration along the coast of southern Alaska and the Pacific Northwest before—or during—the Lateglacial involved moving from one ice-free refugium to another. The southern coast of the land bridge, however, remained unglaciated throughout the last glacial (MIS 2 age equivalent) (Hopkins 1967c), and recent work has revealed the existence of more extensive ice-free zones and early deglaciation of some areas on the Northwest coast (e.g., Mann and Hamilton 1995; Josenshans et al. 1997). Nevertheless, the model still assumes intensive use of watercraft by the migrating population (Dixon 2001:292). It also assumes an economic emphasis on marine resources, including shellfish, ocean fish, and marine mammals (Dixon 1999:249–251), although some coastal terrestrial resources also were available during Lateglacial times (Fladmark 1979:59; Dixon et al. 1997:698–700; Dixon 2001:291). Dates of as early as 17–16 cal ka are suggested for the initial occupation of the south coast of the Bering Land Bridge (Dixon 2001:292).

The primary advantage of the Pacific coast migration model is that it accounts for a human presence south of the ice before 15–14 cal ka, which is the earliest estimated date for the opening of the ice-free corridor (Arnold 2002; Dyke 2004:411). Indeed, if the dates for the Lateglacial corridor continue to fall below those of the oldest documented occupations, a coastal migration becomes necessary to account for the latter (assuming that the transatlantic migration is improbable). The early Paleoindian complexes in the North American interior (e.g., the Clovis complex) presumably are derived from a population of coastal dwellers that subsequently moved inland and developed interior foraging economies (Dixon 1999:247).

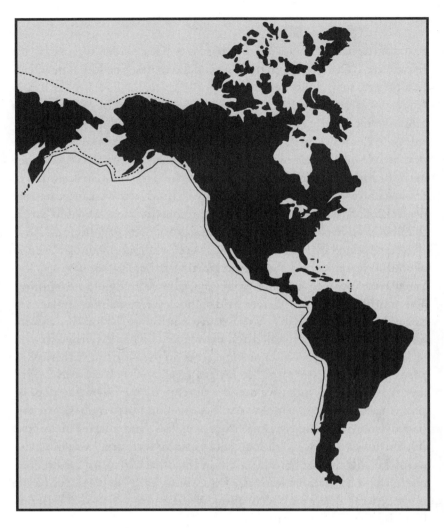

Figure 7.3 The Pacific Coast Migration model of New World settlement

The most serious deficiency of this model is the lack of evidence for marine adaptations in Northeast Asia and along the North Pacific rim before the Holocene (Yesner 2001:316–317; Workman 2001), although most of the sites that would contain such evidence would have been inundated at the end of the Pleistocene. Dixon (2001:292) suggested that a North Pacific marine economy could have initially developed along the southern coast of the Bering Land Bridge (see also Laughlin 1967) instead of in Northeast Asia. Critics of the model also have questioned the availability of food resources and suitable materials for boats along the North Pacific rim during the late Pleistocene (Workman 2001).

The lack of any obvious source in the interior of Beringia for the early Paleoindian complexes of the North American Plains provides important negative supporting evidence for the coastal migration model. Beyond this, the Beringian archaeological record—including southeast Alaska—yields some indirect evidence for settlement of the coastal zone before the Younger Dryas. Although the oldest dated occupations along the coast of southeast Alaska do not antedate the Younger Dryas, the stable isotope values for the human skeletal remains from On-Your-Knees Cave dating to 10 cal ka indicate a heavy marine diet, suggesting that the population had been established on the coast for an extended period (Dixon 2001:286–287). Furthermore, there is some evidence for an earlier industry containing bifaces that antedates the microblade assemblages (Dixon 1999:178–181; Mandryk et al. 2001:308–309), although its age remains unclear.

A majority of North American archaeologists probably favor the coastal migration model at the present time. Because archaeological sites on the southern coast of Alaska (or farther south on the Northwest coast) have yet to be dated to more than 12 cal ka, support for the model rests primarily on the presence of sites south of the ice that antedate the ice-free corridor (Fiedel 2002:421). While some archaeologists question the validity of dates on Monte Verde and other sites greater than 14–13 cal ka (e.g., Fiedel 1999:107), most accept the premise that the settlement of North and South America preceded a viable interior route. An early coastal migration does not preclude interior movement of people from Beringia to the northern Plains after 13.5 cal ka, however, and both models may ultimately prove correct (but see the following discussion).

In the future, archaeologists probably will be able to test the coastal migration hypothesis more directly (Easton 1992). Isostatic rebound has uplifted some late Pleistocene coastal margins to elevations above modern sea level, thus providing opportunities to search for archaeological sites along former shorelines (e.g., Josenhans et al. 1997:74). Developments in remote-sensing technology eventually may permit more effective prospecting for underwater remains.

INTERIOR ROUTE MODEL: ICE-FREE CORRIDOR

If the Pacific coastal migration model enjoys increasingly widespread support among archaeologists, the ice-free corridor route has been rejected by many—at least as the means by which the initial peopling of the New World took place. Sometimes labeled as the "classic model," an interior route is commonly postulated in the earlier archaeological literature (e.g., Wormington 1957:249–260; Wilmsen 1964; Willey 1966:33–37; Haynes 1982:395–398).

In its fundamental form, the classic model envisions a small and highly mobile population of interior foragers crossing the Bering Land Bridge into eastern Beringia to the northern Plains via the ice-free corridor. West (1981:205–206) and others have suggested that similar environments in the Beringian interior, ice-free corridor, and northern Plains facilitated rapid spread of the migrating foragers. This suggests that there might be relatively little difference between Beringian and early Paleoindian technology (Hoffecker, Powers, and Goebel 1993; West 1996a:553–554), although there is near consensus that the diagnostic Clovis-type fluted point was confined to mid-latitude North America.

Leaving aside the problem of sites south of the Canadian ice that seem to antedate a viable ice-free corridor, the interior model still is afflicted with several major problems. The most serious problem is the absence of any known sites in the ice-free corridor zone that may be firmly dated to 13.5–12.5 cal ka (Wilson and Burns 1999) and any sites or isolated remains that appear to be related to the Lateglacial Beringian industry (described in chapters 4 and 5). The earliest dated sites in the ice-free corridor date to the Younger Dryas or later and contain assemblages that suggest close links with the Plains (i.e., from the south) (e.g., Beaudoin, Wright, and Ronaghan 1996). Because the landscape is accessible to archaeologists and has been subject to some survey, this negative evidence carries more weight than the lack of recorded sites along the submerged Pacific coast.

Another issue is the environmental contrast between Lateglacial Beringia and the North American Plains. The mesic shrub tundra that spread across Beringia after 15–14 cal ka contained a postglacial flora and fauna that was very different from that of the northern Plains. Significant contrasts are apparent in the Lateglacial economies and settlement strategies of the two regions, and there is no reason to assume that the technologies would have been similar.

Despite these problems, an interior route retains some advantages as an explanatory framework and is still embraced by some archaeologists (e.g., West 1996a:553–554; Fiedel 1999:108–109; Yesner et al. 2004). One of these is the fact that only the Beringian interior contains well-dated sites that precede the earliest Paleoindian complexes on the Plains and provide a potential source for them. Another advantage is that dates on the earliest Paleoindian sites of the Plains coincide with—or slightly postdate—the opening of the ice-free corridor (Holliday 2000:264–265; Mandryk et al. 2001; Dyke 2004:411).

The Beringian archaeological record indicates that people were present in the eastern interior at least a thousand years before the appearance of the Clovis complex in mid-latitude North America. Although the earliest sites in the Beringian interior cannot account for the occupation of Monte Verde at 15 cal ka, they could represent a source for the early Paleoindian sites on the Plains (Hoffecker 2001:150). There are clearly differences between the Lateglacial Beringian and Paleoindian assemblages, but the degree of perceived difference varies significantly according to the perspective of the investigator. Archaeologists who emphasize the microblade technology—and combine the various Lateglacial assemblages into a single industry—tend to conclude that the differences are substantial (e.g., Holmes 2001:167). West (1996a:554) combines all the Lateglacial (and Younger Dryas) assemblages into one complex (i.e., Denali) but minimizes the differences between them and the Clovis artifacts. Those who believe that more than one complex is present during the Lateglacial often perceive similarities between at least some of the Beringian assemblages and those of the early Paleoindian sites (e.g., Goebel, Powers, and Bigelow 1991; Kunz, Bever, and Adkins 2003).

The ice-free corridor model may eventually enjoy some resurgence of support, but this is likely to depend on several factors. First and foremost, archaeological remains of Lateglacial age that exhibit some link to the Beringian industry and indicate a southward movement at that time must be documented in the former corridor zone. The technological differences between the Beringian and early Paleoindian industries may ultimately be explicable in terms of the different economies and habitats of the two regions. The status of the interior model also depends to some extent on the fate of the coastal migration model. If the evidence for settlement south of the Canadian ice before 14 cal ka is strengthened by new discoveries and dates—and if there is progress in documenting a Lateglacial marine adaptation along the Northwest coast—the interior model probably will be abandoned by most archaeologists.

The Smell of the Shrub Tundra: Rethinking the Archaeology of Beringia

Many archaeologists working on both sides of the Bering Strait have tended to view the archaeological record of Beringia from the perspective of the various models of New World settlement described above. Artifact types and assemblages often have been classified and interpreted in terms of their perceived similarities or differences with those of the Pacific Northwest, Great Plains, or other regions of North America (e.g., Dikov 1979:47–51; Goebel, Powers, and Bigelow 1991; Kunz and Reanier 1994:660; West 1996a:554, fig. 12-2; Dixon 1999:250–253; Adovasio and Pedler 2004).

While the search for links between the Beringian sites and those of both North America and Northeast Asia is a legitimate enterprise—and many of the perceived links are probably valid ones—it has tended to obscure the true character of the Beringian record. The people who settled Beringia during the Lateglacial eventually produced a unique and probably indigenous industry with little resemblance to those of adjoining regions. Although this industry reveals some ties to Northeast Asia—most notably the microblade technology—much of it lacks any obvious links to industries outside Beringia.

The Lateglacial archaeology of Beringia—to paraphrase Jørgen Melgaard's comment about early Dorset culture[9]—yields the "smell of the shrub tundra." Despite the continued presence of a vast plain between Chukotka and Alaska, mesic shrub tundra had spread across Beringia at 15–14 cal ka and apparently rendered it habitable for humans. From the outset, the economy seems to have had more in common with that of later northern interior peoples than of the inhabitants of the late Pleistocene "mammoth steppe." The location, size, and contents of the sites reflect a broad-based diet of postglacial large mammals, small mammals and birds, and some fish. The Beringian economy seems to have had roots in the evolving post-LGM economy of the Siberian interior, but it probably developed in other directions (e.g., greater emphasis on waterfowl).

The most recent discoveries and radiocarbon dates from the lowest level at Swan Point in central Alaska reveal that people making artifacts very similar to those found in the Lena Basin were the first known occupants of Beringia after the LGM (C. E. Holmes, pers. comm., 2006). Their wedge-shaped microblade cores, microblades, and burins are almost identical to artifacts found at Dyuktai Cave on the Aldan River at 15–14 cal ka (Mochanov 1977). A similar assemblage also may be present at Berelekh in northwestern Beringia at this time, although both the dating and contents of Berelekh remain uncertain. The early Lateglacial industry of Beringia is thus linked closely to the late Upper

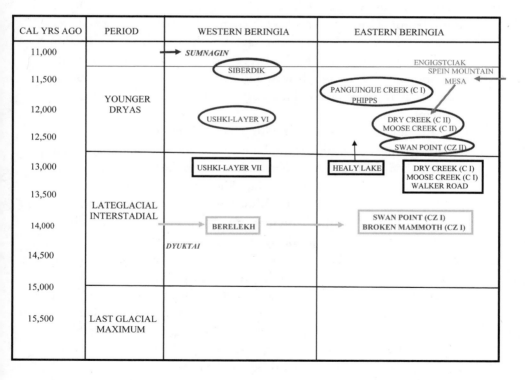

Figure 7.4 Summary of Beringian archaeology as interpreted by the authors. Ushki-Layer VII and other contemporaneous sites enclosed in boxes (dark lines) are assigned to the Beringian tradition.

Paleolithic industry of interior Northeast Asia and could be classified as part of the Dyuktai culture (figure 7.4).

It is during the later centuries of the Lateglacial interstadial (13.5–12.8 cal ka) that the Beringian archaeological record acquires its distinctly native character. Especially diagnostic among the lithic assemblages are a series of small points made on flakes (and often not fully flaked on both sides) that range from stemmed forms at Ushki to triangular and teardrop-shaped forms (Chindadn points) in central Alaska and perhaps the southern Yukon. Yet more variability may be present in northern Alaska at places like Tuluaq Hill.

We concur with Frederick Hadleigh West's (1981:xv) proposal that the assemblages in Beringian sites of the final Lateglacial should be placed under the classification of the Beringian tradition (West 1996a:549–552). It underscores the unique regional character of the Beringian industry that antedates the beginning of the Younger Dryas. A number of complexes have been recognized in this time range (e.g., Chindadn, Nenana, Denali, Sluiceway), but there are many uncertainties about how they fit into the larger entity—activity subsets, post-depositionally mixed assemblages, and so forth. At present, we do not believe that any specific complexes can be identified within this group of assemblages.

At the same time, we concur with Charles E. Holmes, who suggests that the Beringian tradition should not be extended to assemblages that postdate the Lateglacial. Although Holmes (2001:156) places sites of the Younger Dryas and its aftermath into the rather neutral designation Transitional Period (13–10.5 cal ka), such ambivalence may be unnecessary. The Dyuktai culture is present in western Beringia (especially at Ushki Layer VI, but also probably at Kheta), while two distinct traditions can be firmly documented in eastern Beringia (Alaska and the Yukon) at this time. They include the American Paleoarctic tradition and the Paleoindian tradition. The former, although defined as a North American entity (Anderson 1968), contains all the diagnostic elements of the Dyuktai culture of Northeast Asia—wedge-shaped microblade cores, multifaceted burins, and biconvex bifaces. The Paleoarctic tradition is represented most substantively by the Denali complex in central Alaska (West 1967), but also by the Akmak and Kobuk complexes of northwestern Alaska (Anderson 1988). Other manifestations include Ground Hog Bay 2 and On-Your-Knees Cave in southeastern Alaska

(Ackerman 1996e; Dixon 1999:117–119) and Trail Creek Caves on the Seward Peninsula (Larsen 1968).

The Paleoindian tradition, originally defined in mid-latitude North America, contains many complexes (e.g., Clovis, Folsom, and Dalton [Wormington 1957]). In Alaska, the Paleoindian tradition is most clearly represented by the recently defined Mesa complex (Kunz and Reanier 1994), which contains various lanceolate projectile point types of the later Paleoindian sites on the High Plains (e.g., Agate Basin and Hell Gap), as well as some fluted forms that seem to be confined to the north (Kunz, Bever, and Adkins 2003). Mesa sites are most common in the Brooks Range but also are present in southwestern and central Alaska (Ackerman 2001). Paleoindian assemblages are found in the Yukon at Engigstciak (originally assigned to the Flint Creek complex [MacNeish 1956:95–97]) and farther south at Charlie Lake Cave in British Columbia (Fladmark, Driver, and Alexander 1988) and localities in Alberta (Beaudoin, Wright, and Ronaghan 1996).

After the beginning of the Younger Dryas—which marks a sharp break with the preceding interstadial period—the unique character of the Beringian tradition gave way to two established traditions ultimately derived from regions outside Beringia. The expansion of steppic habitat and bison herds apparently created a highway for bison hunters of the High Plains to move northward through the ice-free corridor into eastern Beringia (MacNeish 1963:101–102; Dumond 2001; Mann et al. 2001). For the first time, there was a clear link between the occupants of sites in Beringia and those of the Plains.

The reasons for the explosion of microblade technology in western and especially eastern Beringia during the Younger Dryas are less clear. If the Paleoarctic tradition is a direct outgrowth of the Beringian tradition, it may reflect local developments that were responses to colder climates and their effects on biota (e.g., Mason, Bowers, and Hopkins 2001:539–540) or even perhaps responses to the intrusion of the Paleoindians. Alternatively, a new movement of people eastward from the Lena Basin might have reintroduced typical Dyuktai forms to Beringia. The rapid spread of sites in Northeast Asia assigned to the Sumnagin culture during the Younger Dryas might have been a factor (Pitul'ko 2001:269–272).

Yet another alternative is that people making wedge-shape cores were moving from southern coastal areas into the interior as rising sea levels flooded the Bering Strait and other shelf margins at this time. Although the presence of people in these areas during the Lateglacial cannot be confirmed yet because they lie beneath the sea, the oldest firmly dated assemblages in the southeast Alaska coastal zone—which date to the Younger Dryas—contain wedge-shaped microblade cores (e.g., Ackerman 1996e:426–429). During the Younger Dryas, landscape, climates, and biota of Beringia were all undergoing rapid and dramatic change, and the archaeology reflects this dynamism.

Notes

Chapter 1. An Introduction to Beringia

1. Commissioned in 1967, the *Discoverer* served for three decades as one of the largest research vessels operated by the National Oceanic and Atmospheric Administration.

2. Throughout this book, dates are given in calibrated radiocarbon years and abbreviated as "cal ka" (thousand calibrated years ago). Dates have been calibrated with the current version of the Cologne Radiocarbon Calibration Programme (Calpal), which is available online (www.calpal.de).

3. At the time the new land bridge chronology was developed, archaeologists thought that the oldest site in North America might be at Bluefish Caves in the northern Yukon, which had yielded dates as early as 18.5–15.5 cal ka (Morlan and Cinq-Mars 1982).

4. Boris Yurtsev is one of the few Beringian researchers to approach the problem of defining Beringia in an explicit and systematic manner (e.g., Yurtsev 1984:131–134).

5. Yurtsev's "Megaberingia" extends beyond the Mackenzie River to include a much larger portion of northwestern Canada (Yurtsev 1984:150–151, fig. 11).

6. The recent discovery of Yana RHS ("Rhinoceros Horn Site") near the mouth of the Yana River suggests that the Verkhoyansk Range was a less significant barrier

to human settlement than previously thought (see chapter 3). The Yana RHS site was occupied roughly 30 cal ka during the final stages of the interstadial correlated with marine isotope stage 3 (MIS 3) (Pitul'ko et al. 2004).

7. The bedrock of the Bering-Chukchi Platform is composed of volcanic and intrusive rock in the northeastern Bering Sea and stratified sedimentary rock in the Chukchi Sea (Creager and McManus 1967:15).

8. With the exception of the Alaska Range and the Chugach/St. Elias Mountains, glaciation during MIS 4 was considerably more extensive in Beringia than during the subsequent MIS 2 glaciation. Paleotemperature estimates for MIS 4 in eastern Beringia indicate cold summers, with the July mean about 6°C below modern levels, based on beetle assemblage data from Alaska and the Yukon. Although winter temperatures are estimated to have been near modern levels, this annual temperature profile would have promoted the growth of mountain glaciers, as colder summer temperatures inhibit summer melting of the snowpack, whereas winter temperatures at these latitudes were always sufficiently low to allow formation of glacial ice throughout the Pleistocene (Glushkova 2001; Elias 2001b; Heiser and Roush 2001).

9. Kuzmin and Keates (2005) recently have challenged the conclusion that much of Siberia was abandoned during the Last Glacial Maximum, arguing that radiocarbon dates from several sites indicate human settlement above latitude 55°N.

10. The re-dating of the inundation of the Bering Strait to after 13 cal ka suggests that the increased moisture promoting the transition to mesic shrub tundra across much of Beringia at 17–15 cal ka was not derived from the flooded land bridge. An alternative source of moisture at the beginning of the Lateglacial warming might have been generated by reduced sea ice cover in the Bering Sea (Cassie et al. 2004).

11. Several archaeologists have suggested that the stone artifacts recovered from Diring Yuriakh are the product of natural processes, such as frost action (e.g., Goebel 1999:210).

12. The Neandertals may have occupied a small cave at 54°N on the Upper Yenisei River during the Last Interglacial climatic optimum (MIS 5e), when climates were warmer than those of the present day (Abramova 1981; Hoffecker and Elias 2003:36). Their maximum cold tolerance, however, may be evident in the Altai region, where they occupied cave and open-air sites at relatively high elevations during periods that were colder than they are at present (Hoffecker 2005a:58–59).

13. Although large mammal bone would have provided a fuel source in areas where trees were absent, adequate starter fuel in the form of woody shrubs might have been lacking at times. This also may have been a factor in the Lateglacial settlement of Beringia (Hoffecker and Elias 2005). The effectiveness of technology for cold protection before 20 cal ka is difficult to assess, but sewn clothing and tents produced during the earlier Upper Paleolithic might not have provided adequate insulation from extreme low temperatures (Hoffecker 2005b).

14. Most archaeologists seem to believe that there are no firmly dated sites in North and South America that antedate the oldest known sites in Beringia and thus imply earlier settlement of the latter. At present, evidence for earlier settlement is confined to indirect sources of information. Several studies of genetics and linguistics among modern Native American populations have concluded that they indicate a relatively early migration to the New World—and by implication Beringia—as early as 30 cal ka (e.g., Nichols 1990; Schurr and Sherry 2004). Other studies have drawn conclusions from the genetic, morphologic, and linguistic data that are more consistent with the dates on known archaeological sites (e.g., Greenberg, Turner, and Zegura 1986; Nettle 1999; Brace et al. 2001; Hey 2005).

15. According to experimental research by Théry-Parisot (2001), bone has a high ignition temperature and is difficult to fire without some wood fuel.

16. The most diagnostic artifact form of the early Paleoindian assemblages in mid-latitude North America is the Clovis-type fluted point (Haynes 1982: 384–393). Fluted points have been turning up in northern Alaska and northwestern Canada since 1948, but typical Clovis forms are absent. Like the nonfluted lanceolate points of Alaska, the northern fluted points appear to postdate the early Paleoindian sites of the mid-latitude regions.

Chapter 2. Beringian Landscapes

1. No one lives on mountaintops in Siberia, and the highland region is too vast and remote to allow the collection of long-term meteorological data.

2. The term "permafrost" signifies regions where the ground is permanently frozen. There is a relatively shallow active layer of soil at the surface that thaws during the summer season, varying in depth from more than a meter to just a few centimeters.

3. The southern border of this type of woodland coincides with the Sea of Okhotsk coast, up to the vicinity of Magadan, and including the northern part of the Koni Peninsula. This Beringian Woodland zone is found only in the milder climates in relatively close proximity to maritime environments, and it gradually changes from being dominated by dwarf Siberian pine (inland sites) to alder (coastal regions).

4. The Baird and Delong Mountains have summits averaging 1,000–1,500 meters in elevation. The summits of the Brooks Range become increasingly higher to the east, reaching an average elevation of 2,400 meters. The highest peak in the range is Mount Isto, at 2,760 meters.

5. The crest of most of the range averages between 2,100 and 2,750 meters in elevation, with a number of peaks exceeding 3,000 meters. The huge massif of

Mount McKinley (Denali), at 6,194 meters, is the highest point in North America.

6. The St. Elias Mountains and the Kenai-Chugach mountain system have the most extensive system of highland and valley glaciers in North America, consisting of the Chugach and St. Elias ice fields. The south-facing slopes of the Chugach Mountains support the largest piedmont glaciers in the world (the Malaspina and Bering Glaciers). These and other glaciers extend to the sea, where the never-ending downhill flow of ice sends huge icebergs into numerous inlets, bays, and fiords.

7. Some parts of Alaska are very cold. For example, many mountaintops in the various ranges are covered in ice and snow throughout the year. The arctic coast never gets much above 10°C, even on the warmest days of summer. In contrast, some parts of Alaska have a climate that is almost as mild as Seattle, and snow rarely stays on the ground for more than a few days. The interior regions have quite warm summers. It is not unusual for July and August temperatures in Fairbanks to reach 32°C.

8. The permafrost table is at or near the ground surface in this region, with an active layer of less than 0.50 meters (Brown et al. 1997). Pingos and other ice-related features such as ice-wedge polygons, oriented lakes, peat ridges, and frost boils are common. Thaw lakes are a major feature of these landscapes, covering 20–50 percent of the land in most regions.

9. Open low scrub occurs along drainages. The only coniferous forest stands occur in the Noatak River Valley, consisting of white spruce (*Picea glauca*) occurring in either pure stands or with balsam poplar. Mesic herbaceous tundra is dominated by tussock-forming sedges. Dwarf scrub communities are dominated by mat-forming species of mountain avens (*Dryas octopetalla*), accompanied by heath-family shrubs. Willow and alder stands grow along the banks of streams.

10. White spruce stands usually grow on well-drained soils, such as on hillsides. Black spruce stands tend to grow on poorly drained sites and on north-facing slopes where permafrost is very close to the surface. Tall scrub thickets grow on alluvial deposits that are subject to periodic flooding. Tall scrub swamps and wet graminoid herbaceous communities are found on the wettest sites. Broadleaf forests also grow on flood plains. Mixed forests consist of closed stands of white spruce and paper birch or white spruce and balsam poplar on the better-drained alluvial soils; poorly drained soils support stands of black spruce and paper birch. Broadleaf forests consist of closed stands of balsam poplar.

11. Peat mounds, barren sand dunes, and volcanic soils support dwarf scrub communities dominated by heath family shrubs. In the southern part of this region, where peat or alluvium accumulation and growing-season temperatures are sufficient, some stands of coniferous forest are found. These stands consist of black and white spruce, with an understory of alders, willows, and birches A

variety of mosses covers the ground. Wet meadows are typically dominated by sedges. Bogs develop where peat mounds and polygonal ridges provide drained substrates for woody plants, such as heath-family shrubs.

12. Mean annual temperature is below 0°C in many regions, and the coldest recorded temperature (–62.8°C) in North America was reported here. TMIN averages –25 to –30°C in the north and central parts of the territory, and –15 to –20°C in the southern regions, except for the higher elevations (figure 2.13) (Wahl et al. 1987). TMAX ranges from 10 to 15°C throughout much of the territory, except for the mountains in the southwest, where TMAX ranges from –5 to 5°C (figure 2.14) (Wahl et al. 1987).

13. Typical plant communities in this zone are dominated by cottongrass (*Eriophorum vaginatum*), sedge (*Carex aquatilis*), arctic willow (*Salix arctica*), arctic heather (*Cassiope tetragona*), arctic dock (*Rumex arcticus*), and arctic avens (*Dryas integrifolia*) (Wiken et al. 1981). Upland areas are covered by cottongrass tussocks with shrubs and heather. The shrubs include Labrador-tea (*Ledum decumbens*), mountain cranberry (*Vaccinium vitis-idaea*), and dwarf birch (*Betula nana*). Lowlands are clothed in sedges and *Sphagnum* mosses.

14. Typical alpine plants include mountain avens, arctic bearberry (*Arctostaphylos alpina*), alpine saxifrage (*Saxifraga caespitosa*), and crustose lichens. High mountain ridges and other exposed localities support patches of moss campion (*Silene acaulis*), mountain avens, and saxifrages. Lower slopes are covered by vegetation that includes ground birch (*Betula glandulosa*), willows, crowberry (*Empetrum nigrum*), and *Hylocomium* mosses (Wiken et al. 1981). Sedge tussock-tundra dominates many of the highlands of the Richardson Mountains (Stanek, Alexander, and Simmons 1981), with *Sphagnum*, *Polytrichum*, and *Hylocomium* mosses growing in hummocky ground, interspersed with low shrubs such as dwarf willow.

15. The typical vegetation in this ecosystem consists of stunted black spruce and larch (*Larix laricina*) with lesser numbers of white spruce. The ground cover is dominated by dwarf birch and willows, heathers, cottongrass, lichens and mosses. In some regions, an open canopied lichen-spruce woodland is prevalent. This vegetation contains black spruce, with lichens as the dominant ground cover (Stanek, Alexander, and Simmons 1981). Paper birch, balsam poplar, and green alder (*Alnus crispa*) are important elements of subalpine forests in the Interior Plains region.

16. McManus and Creager (1984) estimated that the land bridge was inundated for the last time at 17.8–17.7 cal ka (14,600–14,400 ^{14}C years BP). Their estimate was based on radiocarbon assays of bulk sediments from terrestrial deposits covered by postglacial marine sands and left on the Chukchi shelf. However, they noted in their article that these sediments may have been contaminated by minute particles of reworked coal. They therefore cautioned their readers to look on this

as a maximum age. Nevertheless, the 14,500 ^{14}C years BP age gained a life of its own in the literature, and for a decade this became the accepted age of inundation of the land bridge.

17. For instance, paleontological evidence suggests that camelids (camels, llamas, alpacas, guanacos, and others) evolved in the Americas during the Tertiary, and that camels migrated west across the Bering Land Bridge into Asia. The two living species of camel are now only found in Asia and North Africa. Several species of North American camel became extinct at the end of the last glaciation. They formed part of the megafaunal grazer community, ranging as far north as eastern Beringia.

18. The soils of the land bridge developed from marine sediments, which were exposed to the air when sea level dropped. Wind-blown loess from adjacent Siberia and Alaska may have contributed to the buildup of terrestrial soils. It is extremely difficult to say what these soils were like since there is nothing to compare them with in the modern world.

19. The question of environmental reconstruction of the Bering Land Bridge is far from being answered. The situation is reminiscent of a team of blind men being asked to describe an elephant (or perhaps a woolly mammoth would be more appropriate). Each one feels only one small part of the elephant and reports findings that are very different from those of each of the others. One feels a leg and believes he is feeling the trunk of a large tree. Another feels the tail and believes he is feeling a small rodent. A third feels the belly and believes he is feeling a very large mammal. None of their data are wrong, but it is difficult to reconcile the different interpretations. The Bering Land Bridge was an enormous region, and paleontologists have thus far analyzed the fossil record from only a few sites.

20. Mesic tundra vegetation at the Toolik Lake Long-Term Ecological Research site on the Alaskan North Slope has a mean net annual primary productivity of 141 g/m^2, although it varies locally from 32 to 305 g/m^2, depending on site characteristics (Shaver and Chapin 1991). Net annual primary productivity in cold desert steppe environments, such as at Shortandy, Kazakhstan (51° 42' N, 71° 00' E) averages 332 g/m^2 (Titlyanova, Kiryushin, and Okhin'ko 1984). This level of productivity is almost as high as that found today in African savannah vegetation. For instance, net annual primary productivity at a biological research station near Nairobi, Kenya, averages 357 g/m^2 (Kinyamario and Imbamba 1992).

21. The only anomalous results in Guthrie's (2001) study concern the Pleistocene horse data, but Guthrie explains this seeming discrepancy between known equid diet and plant remains by showing that the identical pattern turns up in modern African zebras. Equids use their lips to feed on tender grasses and their incisors to cut more resistant (woody) plant material, so a disproportionate amount of woody material becomes lodged in their incisors. The crowns of

equid molars are filled with cementum. They do not trap plant remains, so fossil incisors must be used for paleobotanical studies. Guthrie's work clearly shows that Beringian megafaunal grazers were feeding mainly on grasses.

Chapter 3. Settlement of Northern Asia

1. The "oceanic effect" and its influence on human settlement in northern Eurasia are described by Clive Gamble (1995:283–284). Clockwise currents in the North Atlantic bring warm and moist air to western Europe, but by the time this air reaches eastern Europe and Siberia it is colder and drier.

2. Comparison of primary productivity between Europe and Siberia above 45°N reveals as much as a 50 percent reduction in the latter, owing primarily to the oceanic effect (Archibold 1995). However, primary productivity in the Pleistocene steppe-tundra environment of Beringia probably was higher than that of modern Siberian tundra. As discussed in chapter 2, the modern net annual primary productivity in cold desert steppe environments of Kazakhstan averages 332 g/m^2 (Titlyanova, Kiryushin, and Okhin'ko 1984) and is almost as high as that found today in the African savannah.

3. Several sites in southern Europe (below latitude 45°N) are dated to 1–0.8 mya (e.g., Atapuerca) but may not be related to the occupation of higher latitudes that is evident by 0.5 mya. This later occupation is accompanied by the appearance of handaxes, which are absent in the southern sites before 0.6 mya (Roebroeks and van Kolfschoten 1995).

4. Warm interglacial intervals are correlated with MIS 13, MIS 11, and MIS 9, and the interglacial correlated with MIS 11 was especially warm and protracted (Roebroeks and van Kolfschoten 1995).

5. The Boxgrove tibia also indicates an estimated body weight of more than 80 kg (Stringer and Trinkaus 1999), which is more typical of living peoples in higher latitudes and may indicate some form of cold adaptation with respect to overall body size (i.e., increased ratio of volume to exposed surface area) (Hoffecker 2005a:37–38).

6. For many years, the earliest confirmed use of controlled fire was attributed to layers dated to 0.4 mya at Zhoukoudian in north China, but recent analysis of the deposits at Locality 1 found no evidence for this (Goldberg et al. 2001).

7. Human occupation antedating 0.3 mya has been documented on the northern slope of the Caucasus Mountains at Treugol'naya Cave and possibly on ancient terraces in the Lower Dnestr Valley in southwest Ukraine (Hoffecker 2002a:43–48).

8. As Slobodin (2001:33–35) notes, the debate over Diring has resurrected discussion of some other "pebble tool" sites reported from Siberia in earlier years—Filimoshki and Ulalinka (Powers 1973).

9. Neandertal skeletal remains are known from the southernmost areas of eastern Europe (Crimea and northern Caucasus), but they are extremely rare on the East European Plain, as well as in Siberia (Hoffecker 2002a). The presence of the Neandertals in the latter areas is inferred from the types of stone artifacts found in local Middle Paleolithic sites, which are characteristic of Neandertal sites in adjoining areas (i.e., Levallois Mousterian).

10. Aiello and Wheeler (2003) argue that Neandertal anatomy provided relatively limited protection from low temperatures in comparison with that of modern humans. Also, while the large volume of the Neandertal cranium follows a general pattern of anatomical features that correlate with latitude/temperature among living humans (Holloway 1985:320–321), African *Homo* populations of the late Middle Pleistocene also exhibit large crania—presumably unrelated to climate.

11. Although the bones of the largest herbivores—mammoth and woolly rhinoceros—are not common in western European sites, new stable isotope data indicate that they may have had an important role in Neandertal diet (Bocherens et al. 2005). In eastern Europe, mammoth remains are abundant in many sites, and rhinoceros is usually represented (Hoffecker 2002a:113–119).

12. Under rare circumstances, wooden implements have been preserved in Neandertal sites (e.g., Abric Romani, Spain [Carbonell and Castro-Curel 1992]), but it is not clear whether they represented a significant advance over the wood technology of earlier European hominins or continued production of tools and weapons made by the latter.

13. Christy G. Turner (pers. comm., 2003), who has undertaken field research at Dvuglazka Cave, believes that the deposits containing the Mousterian artifacts are heavily disturbed and that the presence of a Neandertal occupation is problematic.

14. The current January mean in Gorno-Altaisk is –16°C, while the minimum recorded temperature is –44°C (Konyukova, Orlova, and Shver 1971:12–13).

15. Syntax—or the mechanics of sentence formation—is the most important and complex component of fully modern human language. Nonsyntactical languages, which include pidgins, convey a comparatively limited amount of information (Bickerton 1990).

16. *Homo sapiens* remains are dated to 100–90 cal ka at Skhul and Qafzeh (Israel) and are associated with some evidence for ornamentation, art, and burial ritual (d'Errico et al. 2003:20–21). They may represent a brief and limited expansion of modern humans out of Africa.

17. At Kostenki on the Don River in Russia, Upper Paleolithic assemblages and isolated teeth assigned to *Homo sapiens* have been recovered from layers beneath the Y5 tephra, which is dated by Ar/Ar to 38–41 cal ka in the Mediterranean region

(Ton-That, Singer, and Paterne 2001). Infrared-stimulated luminescence (IRSL) dates on sediment underlying the tephra layer at Kostenki have yielded dates of 41–50 cal ka (S. L. Forman, pers. comm., 2002).

18. The oldest modern human sites in western Europe—the warmest and most productive area of northern Eurasia—are no more than 45 cal ka (Mellars 1996:405–411; Bocquet-Appel and Demars 2000). The dispersal of modern humans into Siberia may have been part of a larger pattern of higher-latitude colonization that initially favored areas where Neandertals were less common or absent altogether. Modern humans perhaps employed their ability to create new technological and organizational structures to occupy environments that were too cold or unproductive to support archaic human populations (Hoffecker 2005a:76–82).

19. Modern human skeletal remains in Siberia dating to 40–20 cal ka are confined to isolated fragments, but they appear to be similar to the west Eurasian population (Turner 1990).

20. Brantingham et al. (2001) suggest that the sequence of Middle and Upper Paleolithic stone artifact assemblages at Kara-Bom indicate a gradual transition, but this seems unlikely, given the apparent replacement of the local archaic population by modern humans at approximately the same time.

21. Tolbaga yielded a bone with markings that may represent the engraved image of a bear's head. This is the only possible reported example of art from Siberia that antedates 30 cal ka (Vasil'ev, Kuznetsov, and Meshcherin 1987). The cultural layer at Tolbaga, like many other Siberian open-air sites, has been subject to significant frost disturbance, and the interpretation of the features may be problematic (Goebel, Waters, and Meshcherin 2001:65; S. A. Vasil'ev, pers. comm., 2004).

22. Hare is reported from Vavarina Gora and Malaya Syya (Abramova 1989).

23. No sites dating to this time range are currently known along the Okhotsk Sea coast, although much of the area is now underwater. Several sites yielding dates of 35,000–25,000 ^{14}C years BP were found on the Aldan River, which might represent a winter area for the occupants of Yana RHS (Mochanov 1977:34–58), but their dates are controversial (e.g., Yi and Clark 1985; Abramova 1989). In both cases, the distances to the Lower Yana River region are considerable (1,000 km or more).

24. The apparent absence of archaeological sites in more southerly regions of the New World dating to the LGM does not necessarily imply an absence of sites in Beringia at this time. The northern portions of the "ice-free corridor" between the Laurentide and Cordilleran ice sheets apparently were blocked by glacial ice as early as 30 cal ka (Mandryk et al. 2001:302–304).

25. This was apparently not the case further north. As noted in chapter 2, LGM climates in Beringia were drier, but not markedly colder, than today. Alfimov and Berman (2001) estimate that mean summer temperatures in western Beringia were

as high as 12–13°C. These temperatures are very close to modern levels in many parts of northeast Asia. Near the Laptev Sea coast, Alfimov, Berman, and Sher (2003) estimate that summer temperatures were considerably higher than they are in the coastal lowlands of northeastern Siberia today due to increased continentality of climate.

26. In their list of Siberian occupations with dates in the 23–21 cal ka time range, Kuzmin and Keates (2005:785, table 3) include several sites located at relatively low latitudes, such as Ust'-Ulma 1 on the Amur River and Ogonki 5 on Sakhalin Island. Other sites on the list include occupations that yielded one or more dates in this time range but are widely assigned to other time periods, such as Novoselovo VI (Vasil'ev 1996:184, fig. 143). When these sites are deleted from the list, a decline in dated occupations above latitude 50°N in the Siberian interior—relative to earlier and later time periods—is evident.

27. Experimental research indicates that bone has a high ignition temperature and requires some wood as a starter fuel (Théry-Parisot 2001).

28. As noted earlier, the modern humans who colonized northern Eurasia 45–20 cal ka were derived from southern latitudes, and the sample of skeletal remains from Europe indicates that they retained many of the anatomical characteristics of people from the tropical zone (e.g., Holliday 1999). Medical research by the U.S. Army found that African-American soldiers, who possess similar body dimensions, were subject to comparatively high rates of cold injury in low-temperature conditions (Orr and Fainer 1952:191; Miller and Bjornson 1962:247; Sumner, Criblez, and Doolittle 1974:456). This may explain why the European population—despite their tailored fur clothing—abandoned areas like the central East European Plain during the cold maximum at 23–21 cal ka (Hoffecker 2002b:130). If the body dimensions of the modern humans in Siberia at this time were similar, they also might have been especially susceptible to cold injury (Hoffecker and Elias 2003:37–38).

29. An ethnographic survey of clothing among arctic peoples noted some significant variations that suggest differences in past development of insulated clothing (Hatt 1969).

Chapter 4. The Beginning of the Lateglacial

1. The artifacts at Kukhtui III are buried in sandy loam deposits on the 25-m terrace of the Kuhtui River. Artifacts in the upper portion of the loam have been dated to the middle Holocene (4700 ± 100 ^{14}C years BP) (Mochanov 1977:87–88).

2. Vereshchagin (1977:49) suggested that mammoths might have died in polynias on the river (i.e., an opening in the ice cover).

3. Mochanov (1977:79–80) originally reported the recovery of a wedge-shaped microblade core from the sediment, but this item appears more likely to be a burin or another type of artifact (e.g., Goebel and Slobodin 1999:121).

4. Through the courtesy of Prof. N.K. Vereshchagin, one of the authors (JFH) had an opportunity in 1991 to examine and photograph this interesting find. Two finely worked bifacial point tips were recovered from the bank (Mochanov 1977:80–81).

5. There are some minor inconsistencies in the numbers of faunal remains reported in association with the artifacts between Vereshchagin (1977:42, table 12) and Mochanov (1977:79).

6. The primary source of uncertainty regarding the age and context of the occupation layer at Berelekh appears to be a report that the artifact layer was deposited at a depth of only 150 cm below the surface and therefore stratigraphically above the mammoth "cemetery" (Vereshchagin and Ukraintseva 1985:111–112). According to Mochanov (1977:78), this was an error: the correct depth was 250 cm.

7. Mochanov (1977:86) noted that some of the mammoth bones were found in a vertical position (although this could be accounted for by frost action). The representation of mammoth body parts (Vereshchagin 1977:18, table 3) is similar to that of sites on the East European Plain that contain remains of mammoth-bone dwellings (Pidoplichko 1969; Soffer 1985).

8. Another concentration of large mammal bones of late Pleistocene age is described from the Lower Kolyma Basin at Bochanut. The precise age of the remains is unclear, but according to Sher, they include mammoth, woolly rhinoceros, horse, bison, and others. Although some of the bones from Bochanut reportedly exhibit traces of human activity (Mochanov 1977:93), this seems to be problematic.

9. Helge Larsen also recovered a small bifacial point fragment of chalcedony from the basal clay layer outside the entrance to Cave 2 (Larsen 1968:56, fig. 37). The association and dating of this artifact have never been clear, but it resembles small bifacial points later found in other parts of Beringia and dating to ca. 13 cal ka (Holmes 2001:164, fig. 7).

10. Thousands of specimens in the Fairbanks muck were collected by Otto W. Geist and others during the 1940s and 1950s and shipped to the American Museum of Natural History for study. In 1980, one of us (JFH) had an opportunity to examine several hundred of them.

11. Robert Ackerman excavated a total of 4 m^2 at Lime Hills Cave 1 during 1993. Radiocarbon dates obtained on charcoal from the upper component were as follows: 8,150 ± 80 ^{14}C years BP (Beta-67668); 8,480 ± 260 ^{14}C years BP (WSU-4504); 8,480 ± 190 ^{14}C years BP (WSU-4505); and 9,530 ± 60 ^{14}C years BP (Beta-67667) (Ackerman 1996a:473, table 10-6).

12. Morlan and Cinq-Mars (1982:359) wrote: "Travelling along the Old Crow River, one cannot help but notice the thousands of fossilized bones, teeth, tusks, and antlers that literally pave many of the point bars."

13. Later investigations by Irving, Jopling, and Beebe (1986) suggested even earlier dates for human occupation of the northern Yukon—as far back as the late Middle Pleistocene.

14. Bones identified as tools included a modified caribou tibia that yielded an AMS radiocarbon date of 24,820 ^{14}C years BP and mammoth bones—thought to represent a core and flake—with a mean AMS radiocarbon age of 23,500 ^{14}C years BP (Cinq-Mars 1990:23–24). According to G. Haynes (1983b:171), many of the Bluefish Caves bones exhibit heavy gnawing damage from small and medium canids ("kennel damage").

Chapter 5. The End of the Lateglacial Insterstadial

1. Dikov followed an established tradition in the Paleolithic archaeology of the Soviet Union by exposing broad horizontal areas and former dwellings and other features on occupation floors (e.g., Hoffecker 2002a:127–131). He made explicit comparisons between the Ushki features and the multichambered structures mapped on Upper Paleolithic floors in eastern Europe such as Kostenki 4 and Pushkari I (e.g., Dikov 1979:32).

2. Although Dikov and Titov (1984:70–71) described the sediments containing the lowermost horizons at the Ushki sites as pyroclastic, Goebel, Waters, and Dikova (2003:505n15) dispute this conclusion. Ted Goebel (pers. comm., 2006) notes that warm-water springs feed the lake and prevent it from freezing during winter. If springs were active in Lateglacial and Younger Dryas times, they probably acted as a major draw to the occupants of the sites—especially during colder months.

3. A "Paleolithic Layer VIII" was originally reported 15 cm below Layer VII, although apparently it yielded only charcoal (Shilo, Dikov, and Lozhkin 1967:34) and was eventually omitted from descriptions of the archaeology (e.g., Dikov 1996:244).

4. Dikov (1996:247) has described the avian gastroliths as derived specifically from ducks, although the basis for the taxonomic identification is unclear (size?).

5. The excavator may have had second thoughts about his interpretation of this feature, which he later characterized as a "possible burial pit" (Dikov 1996:247).

6. Dikov (1967:23–30) originally reported the recovery of several microblades in Layer VII but later classified these objects as burin spalls (Dikov 1979:33).

Goebel, Waters, and Dikova (2003:503) did not find any microblades in the Layer VII sample (*n* = 332) recovered from Ushki V in 2000.

7. Goebel, Waters, and Dikova (2003:503) recently described them as small, thin, and "more like notched points than stemmed points."

8. According to Dikov (1996:247), at least some of the excavated hearths in Layer VII exhibited alternating layers of burned charcoal and gray and yellow calcined bone in profile.

9. At each site, the lowermost occupation levels are separated from the overlying occupation levels by 20 cm or more of sterile sediment (Holmes 2001:158, fig. 2). The younger levels yield radiocarbon ages of 12.7–11.1 cal ka.

10. Cook (1969:277–286) originally identified four cultural stages among the sequence of ten levels: Athapaskan (levels 1–2), Tuktu (level 3), Quartzite (levels 4–5), and Early (levels 6–9).

11. Most microblades from the Chindadn horizon (90 percent) were recovered from the upper two levels (levels 6–7), while no microblades are reported from the lower two levels (levels 9–10) (Cook 1996:326, table 6-7).

12. Equally as important as the artifacts, but less widely noticed from the Healy Lake excavations at the time, was the evidence for an emphasis on smaller game—specifically birds—and the heavy use of bone for fuel in the Chindadn levels (e.g., Cook 1969:277–278). Both now have been recognized as significant patterns in the archaeological record of Beringia, but it is apparent that they were documented at Healy Lake more than three decades ago.

13. Archaeological remains had been reported from Moose Creek Bluff as early as 1941 by J. Louis Giddings in the form of pictographs (Lively 1996:309).

14. The Chugwater site has sometimes been referred to as "Moose Creek Bluff," but this has caused occasional confusion with the site of Moose Creek in the Nenana Valley.

15. Although no faunal remains identified to species are reported from Chugwater, R. A. Lively (1996:310) notes that a few bone fragments—assigned to birds and small mammals—and some avian gastroliths were found in test excavations. It is not clear whether these remains are associated with either the lower or the middle component.

16. Geoarchaeological surveys in the North Alaska Range attempted to find older loess that might contain earlier occupation levels (Hoffecker 1988), but such deposits were never discovered in the Nenana Valley where surfaces throughout were apparently barren before the Lateglacial due to high wind velocities or a lack of adequate sediment source, or both (Thorson and Bender 1985).

17. Dung may not have been the only alternative to bone fuel used at Dry Creek. According to Thorson and Hamilton (1977:165–167), several of the radiocarbon samples appear to have been contaminated by local deposits of Tertiary lignite (Wahrhaftig 1958). Although the contamination might have been

introduced by airborne particles, an alternative possibility is that the occupants of Dry Creek were burning lignite in their hearths. Because lignite has a very high ignition temperature of 500°C (Théry-Parisot 2001:136, fig. 38), it would have required either wood or bone tinder.

18. Artifacts from the basal loess unit at Moose Creek had been classified as Nenana complex in the 1980s (Powers and Hoffecker 1989:278), but the similarity of some of these artifacts to those found in younger Denali complex assemblages led to a reassessment (Hoffecker 1996:366). The excavations conducted by Georges A. Pearson in 1996 revealed the presence of upper and lower components in the basal loess, which were classified as Denali and Nenana, respectively (Pearson 1999).

Chapter 6. The Younger Dryas and the End of Beringia

1. Guthrie (1990:285–286) suggests that the rise in bison numbers that occurred at this time was due in part to reduced competition with other large grazers that had become extinct.

2. Dikov (1977:58) also reported a radiocarbon date of 25.2 cal ka on Layer VI at Ushki I (see table 6.1 in this volume), but the sample was collected from another context.

3. Dikov (1996:245) notes that the former structures in Layer VI at Ushki I may represent the largest known concentration of Paleolithic dwellings in the world. And some of the structures in Layer VI contain the earliest confirmed traces of entrance tunnels (Hoffecker 2005a:117).

4. The domesticated dog at Ushki represents the oldest known specimen from Northeast Asia and Beringia, but not northern Eurasia. Dog remains have been identified at 18 cal ka from Eliseevichi in Russia (Sablin and Khlopachev 2002).

5. The contrast in the number of sites dated to the Younger Dryas period in western and eastern Beringia cannot be fully accounted for by differences in recent development. Although many of the Alaskan sites in this time range are found along highway corridors in the Tanana Basin (e.g., Chugwater, Broken Mammoth, and Healy Lake) and valleys of the North Alaska Range (e.g., Dry Creek and Gerstle River), others are found in remote areas of northern Alaska and the Yukon (including Engigstciak, Mesa, and Bedwell), as well as inaccessible parts of the southern interior (e.g., Spein Mountain).

6. The estimated age of Cultural Layer III at Siberdik is based on six radiocarbon dates on charcoal ranging between 9,700 ± 500 ^{14}C years BP and 7,080 ± 600 ^{14}C years BP (Goebel and Slobodin 1999:108, table I) and yielding a mean of 8,212 ^{14}C years BP, or slightly older than 9 cal ka. Two other dates lie outside this range and were excluded: 4,570 ± 370 ^{14}C years BP and 13,225 ± 230 ^{14}C years BP.

7. The traces of soil formation in the eolian sand may correspond to a widely recognized buried soil horizon in Beringia at 11–10 cal ka. The sand, in turn, might have been deposited during the Younger Dryas.

8. The Utukok River fluted point was discovered by Edward Sable of the U.S. Geological Survey and was reported by Raymond Thompson (Thompson 1948; Solecki 1996:513).

9. Kunz, Bever, and Adkins (2003:38) suggest that the microblade assemblage at Mesa is probably related to the assemblage at the Lisburne site, which is located roughly 10 km north of Mesa and dated to roughly 3.8 cal ka.

10. West (1967:371–372) also included the Campus site in the Tanana Valley as representative of the Denali complex.

11. Although the generic identification of *Bison* is unambiguous, the specific assignment (*B. priscus*) remains tentative (Guthrie 1983a).

12. Concentration H contained a single microblade fragment, and M yielded a microblade core. In both cases, these items seem to have been derived from elsewhere and probably are unrelated to the activities represented by the remaining items in the concentration (Hoffecker 1983).

13. West, Robinson, and Curran (1996:384) also report four gravers among the formal tools in the Phipps assemblage. It should be noted that these artifacts were made on snapped flakes and lack the spur that is characteristic of gravers found in assemblages assigned to the Mesa complex (Reanier 1996:506, fig. 11-9; Kunz, Bever, and Adkins 2003:31–32).

14. The current sample of Lateglacial assemblages would be increased by new discoveries but could also be augmented by more precise dating of some existing assemblages (e.g., Bluefish Caves, Chugwater) (see chapter 5).

15. A unifacially fluted point, along with other unfluted bifacial points and microblades, was recovered from the Uptar site near Magadan on the coast of the Okhotsk Sea, but it is dated to about 9 cal ka and seems unrelated to the Paleo-indian complex on eastern Beringia (King and Slobodin 1996).

16. An assemblage from the Little John site in the southern Yukon (described in chapter 5) yielded two AMS dates in the post–Younger Dryas time range, but it is tentatively assigned to the Denali complex. It contains remains of caribou, hare, swan, and small birds (N. Easton, pers. comm., 2006).

Chapter 7. Beringia and the New World

1. As noted in chapter 5, Bolshoi El'gakhchan I on the Omolon River in Chukotka yielded artifacts similar to those found in Layer VII at Ushki (Kir'yak 1993:19–23), but these remain undated and may represent a younger and unrelated industry.

2. Describing the Alaskan Kutchin, Richard Nelson (1973:142) reported, "Hares were an important source of meat . . . . Some families used them for a good percentage of their meals during certain periods, such as midwinter . . . . A few families probably consumed more hare meat during 1969–70 than any other food, largely because other game species such as fish, moose, and bear were scarce at the time."

3. There is no basis for inferring the presence or absence of transportation technology such as boats or sleds during the Lateglacial, but it may be noted that a wooden sledge-runner fragment is reported from a site on Zhokhov Island (north of Berelekh) dated to 9.5–9 cal ka and containing microblade technology (assigned to the early Holocene Sumnagin culture) (Pitul'ko 1993).

4. Although priority in naming the lanceolate point complex of eastern Beringia might be accorded to Richard MacNeish's (1956:95–96) Flint Creek complex, the assemblage at Engigstciak contains microblades.

5. The assemblage recovered from Cultural Zone 3 at Broken Mammoth contains a basally concave point fragment but lacks microblades (Holmes 1996)—it may represent a Mesa complex occupation or a situation similar to Component II at Dry Creek, where both Mesa and Denali assemblages co-occur on the same level. The faunal assemblage from this occupation indicates a broad-based diet and economy (see table 6.3 in this volume).

6. Most archaeologists who have addressed the problem of New World origins have not even seriously considered the possibility of a direct connection with the Upper Paleolithic peoples of Europe.

7. A severe critic of the transatlantic model, Straus (2000:221–222) argues that the similarities between Solutrean and Clovis lithic technology are exaggerated.

8. Dixon (1999:251–253) suggests that the detachable foreshaft with attached stone point thought to have been produced by early Paleoindians might have been derived from a detachable marine mammal-hunting harpoon.

9. In a frequently quoted passage, Jørgen Melgaard (1962:95) suggested that some traits found in early Dorset culture "smell of (the) forest."

Bibliography

Abramova, Z.A. 1979. *Paleolit Eniseya: Kokorevskaya kul'tura.* Novosibirsk: Nauka.

———. 1981. Must'erskii grot Dvuglazka v Khakasii (predvaritel'noe soobshsche-nie). *Kratkie Soobshcheniya Instituta Arkheologii* 165:74–78.

———. 1989. Paleolit Severnoi Azii. In P.I. Boriskovskii, ed., *Paleolit Kavkaza i Severnoi Azii,* 145–243. Leningrad: Nauka.

Abramova, Z.A., S.N. Astakhov, S.A. Vasil'ev, N.M. Ermolova, and N.F. Lisit-syn. 1991. *Paleolit Eniseya.* Leningrad: Nauka.

Ackerman, R. E. 1996a. Cave 1, Lime Hills. In F. H. West, ed., *American begin-nings: The prehistory and paleoecology of Beringia,* 470–475. Chicago: University of Chicago Press.

———. 1996b. Spein Mountain. In F. H. West, ed., *American beginnings: The pre-history and paleoecology of Beringia,* 456–460. Chicago: University of Chicago Press.

———. 1996c. Nukluk Mountain. In F. H. West, ed., *American beginnings: The prehistory and paleoecology of Beringia,* 461–464. Chicago: University of Chicago Press.

———. 1996d. Ilnuk site. In F. H. West, ed., *American beginnings: The prehistory and paleoecology of Beringia,* 464–470. Chicago: University of Chicago Press.

———. 1996e. Ground Hog Bay, site 2. In F. H. West, ed., *American beginnings: The prehistory and paleoecology of Beringia*, 424–430. Chicago: University of Chicago Press.

———. 2001. Spein Mountain: A Mesa complex site in southwestern Alaska. *Arctic Anthropology* 38(2):81–97.

Adams, J. M., and H. Faure. 1997. Preliminary vegetation maps of the world since the Last Glacial Maximum: An aid to archaeological understanding. *Journal of Archaeological Science* 24:623–647.

Adovasio, J. M., and D. R. Pedler. 2004. Pre-Clovis sites and their implications for human occupation before the Last Glacial Maximum. In D. B. Madsen, ed., *Entering North America: Northeast Asia and Beringia before the Last Glacial Maximum*, 139–158. Salt Lake City: University of Utah Press.

Ager, T. A. 1975. *Late Quaternary environmental history of the Tanana Valley, Alaska.* Report 54. Athens: Ohio State University, Institute of Polar Studies.

———. 1983. Holocene vegetational history of Alaska. In H. E. Wright, ed., *Late Quaternary environments of the United States.* Vol. 1: *The Holocene*, 128–141. Minneapolis: University of Minnesota Press.

———. 2003. Late Quaternary vegetation and climate history of the central Bering land bridge from St. Michael Island, western Alaska. *Quaternary Research* 60:19–32.

Aiello, L. C., and P. Wheeler. 2003. Neanderthal thermoregulation and the glacial climate. In T. H. van Andel and W. Davies, eds., *Neanderthals and modern humans in the European landscape of the last glaciation: Archaeological results of the Stage 3 Project*, 147–166. Cambridge, U.K.: McDonald Institute for Archaeological Research.

Alekseev, V. P. 1998. The physical specificities of Paleolithic hominids in Siberia. In A. P. Derevianko, ed., *The Paleolithic of Siberia: New discoveries and interpretations*, 329–335. Urbana: University of Illinois Press.

Alexander, H. L. 1987. *Putu: A fluted point site in Alaska.* Publication No. 17. Burnaby, B.C.: Department of Anthropology, Simon Fraser University.

Alfimov, A. V., and D. I. Berman. 2001. Beringian climate during the Late Pleistocene and Holocene. *Quaternary Science Reviews* 20:127–134.

Alfimov, A. V., D. I. Berman, and A. V. Sher. 2003. Tundrostepnye gruppirovki nasekomykh i rekonstruktsiya klimata pozdnego pleistotsena nizovii Kolymy. *Zoologicheskii zhurnal* 82:281–300.

Alroy, J. 2001. A multispecies overkill simulation of the end-Pleistocene megafaunal mass extinction. *Science* 292:1893–1896.

Anderson, D. D. 1968. A Stone Age campsite at the gateway to America. *Scientific American* 218(6):24–33.

———. 1970. Akmak: An early archaeological assemblage from Onion Portage, Northwest Alaska. *Acta Arctica* 16.

———. 1988. Onion Portage: The archaeology of a stratified site from the Kobuk River, Northwest Alaska. *Anthropological Papers of the University of Alaska* 22(1–2).

Anderson, E. 1984. Who's who in the Pleistocene: A mammalian beastiary. In P. S. Martin and R. G. Klein, eds., *Quaternary extinctions, a prehistoric revolution*, 40–89. Tucson: University of Arizona Press.

Anderson, P. M., P. J. Bartlein, and L. B. Brubaker. 1994. Late Quaternary history of tundra vegetation in northwestern Alaska. *Quaternary Research* 41:306–315.

Anderson, P. M., and L. B. Brubaker. 1994. Vegetation history of northcentral Alaska: A mapped summary of late-Quaternary pollen data. *Quaternary Science Reviews* 13:71–92.

Anderson, P. M., and A. V. Lozhkin. 2001. The Stage 3 interstadial complex (Karginskii/Middle Wisconsinan interval) of Beringia: Variations in paleoenvironments and implications for paleoclimatic interpretations. *Quaternary Science Reviews* 20:93–125.

———. 2002. *Late Quaternary vegetation and climate of Siberia and the Russian Far East (Palynological and Radiocarbon Database)*. Washington, D.C.: U.S. National Oceanic and Atmospheric Administration Paleoclimatology Program; and Magadan: Russian Academy of Sciences, Far East Branch, North East Science Center.

Anderson, P. M., A. V. Lozhkin, B. V. Beleya, O. Y. Glushkova, and L. B. Brubaker. 1997. A lacustrine pollen record from near altitudinal forest limit, upper Kolyma Region, northeastern Siberia. *Holocene* 7:331–335.

Anderson, S., and F. Ertug-Yaras. 1998. Fuel, fodder, and faeces: An ethnographic and botanical study of dung fuel use in central Anatolia. *Environmental Archaeology* 1:99–109.

Anderson-Gerfaud, P. 1990. Aspects of behaviour in the middle Palaeolithic: Functional analysis of stone tools from southwest France. In P. Mellars, ed., *The emergence of modern humans*, 389–418. Edinburgh: Edinburgh University Press.

Anikovich, M. V. 2003. The early Upper Paleolithic in Eastern Europe. *Archaeology, Ethnology and Anthropology of Eurasia* 2(14):15–29.

Archibold, O. W. 1995. *Ecology of world vegetation*. London: Chapman and Hall.

Arnold, T. G. 2002. Radiocarbon dates from the ice-free corridor. *Radiocarbon* 44(2):437–454.

Bader, O. N., and V. E. Flint. 1977. Gravirovka na bivne mamonta s Berelekha. *Trudy Zoologicheskogo instituta AN SSSR* 72:68–72.

Barclay, M., A. Malmin, and D. Croes. 2005. *Hoko River archeological site complex digital image archive*. Pullman: Washington State University Press.

Bard, E., B. Hamelin, R. G. Fairbanks, and A. Zindler. 1990. Calibration of the ^{14}C timescale over the past 30,000 years using mass spectrometric U-Th ages from Barbados corals. *Nature* 345:405–410.

Barnowsky, A. D., P. L. Koch, R. S. Feranec, S. L. Wing, and A. B. Shabel. 2004. Assessing the causes of late Pleistocene extinctions on the continents. *Science* 306:70–75.

Bar-Yosef, O., and A. Belfer-Cohen. 2001. From Africa to Eurasia: Early dispersals. *Quaternary International* 75:19–28.

Baryshnikov, G., J. F. Hoffecker, and R. L. Burgess. 1996. Zooarchaeology and palaeontology of Mezmaiskaya Cave, northwestern Caucasus. *Journal of Archaeological Science* 23:313–335.

Beaudoin, A. B., M. Wright, and B. Ronaghan. 1996. Late Quaternary landscape history and archaeology in the "ice-free corridor": Some recent results from Alberta. *Quaternary International* 32:113–126.

Benecke, N. 1987. Studies on early dog remains from northern Europe. *Journal of Archaeological Science* 14:31–49.

Bever, M. R. 2001. Stone tool technology and the Mesa complex: Developing a framework of Alaskan Paleoindian prehistory. *Arctic Anthropology* 38(2): 98–118.

Bickerton, D. 1990. *Language and species.* Chicago: University of Chicago Press.

Bigelow, N. H., and M. E. Edwards. 2001. A 14,000 yr paleoenvironmental record from Windmill Lake, central Alaska: Late-glacial and Holocene vegetation in the Alaska Range. *Quaternary Science Reviews* 20:203–215.

Bigelow, N. H., and W. R. Powers. 2001. Climate, vegetation, and archaeology 14,000–9000 cal yr B.P. in central Alaska. *Arctic Anthropology* 38(2):171–195.

Bliss, L. C., and J. H. Richards. 1982. Present-day arctic vegetation and ecosystems as a predictive tool for the arctic-steppe mammoth biome. In D. M. Hopkins, J. V. Matthews Jr., C. E. Schweger, and S. B. Young, eds., *Paleoecology of Beringia,* 241–257. New York: Academic Press.

Boaz, N. T., and R. L. Ciochon. 2004. *Dragon Bone Hill: An Ice-Age saga of* Homo erectus. Oxford: Oxford University Press.

Bocherens, H., D. Billiou, A. Mariotti, M. Patou-Mathis, M. Otte, D. Bonjean, and M. Toussaint. 1999. Palaeoenvironmental and palaeodietary implications of isotopic biogeochemistry of Last Interglacial Neanderthal and mammal bones at Scladina Cave (Belgium). *Journal of Archaeological Science* 26:599–607.

Bocherens, H., D. G. Drucker, D. Billiou, M. Patou-Mathis, and B. Vandermeersch. 2005. Isotopic evidence for diet and subsistence pattern of the Saint-Cesaire I Neanderthal: Review and use of a multi-source mixing model. *Journal of Human Evolution* 49:71–87.

Bocquet-Appel, J.-P., and P. Y. Demars. 2000. Neanderthal contraction and modern human colonization of Europe. *Antiquity* 74:544–552.

Bonatto, S. L., and F. M. Salzano. 1997. Diversity and age of the four major mtDNA haplogroups, and their implications for the peopling of the New World. *American Journal of Human Genetics* 61:1413–1423.

Bonnichsen, R. 1978. Critical arguments for Pleistocene artifacts from the Old Crow Basin, Yukon: A preliminary statement. In A. L. Bryan, ed., *Early Man in America from a circum-Pacific perspective,* 102–118. Occasional Paper No. 1. Calgary: Department of Anthropology, University of Alberta.

Boorstin, D. J. 1983. *The discoverers.* New York: Random House.

Bowers, P. M., C. M. Williams, R. C. Betts, O. K. Mason, R. T. Gould, and M. L. Moss. 1996. The North Point site: Archeological investigations of a prehistoric wet site at Port Houghton, Alaska. Report submitted to USDA Forest Service, Tongass National Forest, Alaska. Prepared by Northern Land Use Research, Inc., Fairbanks, Alaska.

Bowler, J. M., H. Johnston, J. M. Olley, J. R. Prescott, R. G. Roberts, W. Shawcross, and N. A. Spooner. 2003. New ages for human occupation and climatic change at Lake Mungo, Australia. *Nature* 421:837–840.

Brace, C. L., A. R. Nelson, N. Seguchi, M. Oe, L. Sering, P. Qifent, L. Yongyi, and D. Tumen. 2001. Old World sources of the first New World human inhabitants: A comparative craniofacial view. *Proceedings of the National Academy of Sciences* 98:10017–10022.

Bradley, B., and D. Stanford. 2004. The North Atlantic ice-edge corridor: A possible Paleolithic route to the New World. *World Archaeology* 36(4):459–578.

Brain, C. K., and A. Sillent. 1988. Evidence from the Swartkrans Cave for the earliest use of fire. *Nature* 336:464–466.

Brantingham, P. J., A. I. Krivoshapkin, L. Jinzeng, and Ya. Tserendagva. 2001. The initial Upper Paleolithic in Northeast Asia. *Current Anthropology* 42(5): 735–747.

Brigham-Grette, J. 2001. New perspectives on Beringian Quaternary paleogeography, stratigraphy, and glacial history. *Quaternary Science Reviews* 20:15–24.

Brigham-Grette, J., and D. M. Hopkins. 1995. Emergent marine record and paleoclimate of the Last Interglaciation along the northwest Alaskan coast. *Quaternary Research* 43:159–173.

Brigham-Grette, J., D. M. Hopkins, S. L. Benson, P. Heiser, V. F. Ivanov, A. Basilyan, and V. Pushkar. 2001. Last Interglacial sea level record and Stage 5 glaciation of Chukotka Peninsula and St. Lawrence Island. *Quaternary Science Reviews* 20:419–436.

Brigham-Grette, J., A. V. Lozhkin, P. M. Anderson, and O. Y. Glushkova. 2004. Paleoenvironmental conditions in western Beringia before and during the Last Glacial Maximum. In D. B. Madsen, ed., *Entering America. Northeast Asia and Beringia before the Last Glacial Maximum,* 29–61. Salt Lake City: University of Utah Press.

Brook, B. W., and D. M. Bowman. 2002. Explaining the Pleistocene megafaunal extinctions: Models, chronologies, and assumptions. *Proceedings of the National Academy of Sciences* 99:14624–14627.

Brown, J., O. J. Ferrians Jr., J. A. Heginbottom, and E. E. Melinkov. 1997. *Circum-Arctic map of permafrost and ground ice conditions*. Circum-Pacific Map Series No. CP-45. Washington, D.C.: U.S. Geological Survey.

Brubaker, L. B., P. M. Anderson, M. E. Edwards, and A. V. Lozhkin. 2005. Beringia as a glacial refugium for boreal trees and shrubs: New perspectives from mapped pollen data. *Journal of Biogeography* 32:833–848.

Brubaker, L. B., P. M. Anderson, and F. S. Hu. 2001. Vegetation ecotone dynamics in southwest Alaska during the late Quaternary. *Quaternary Science Reviews* 20:175–188.

Cachel, S., and J. W. K. Harris. 1998. The lifeways of *Homo erectus* inferred from archaeology and evolutionary ecology: A perspective from East Africa. In M. D. Petraglia and R. Korisetter, eds., *Early human behaviour in global context: The rise and diversity of the Lower Palaeolithic record*, 108–132. London: Routledge.

Carbonell, E., and Z. Castro-Curel. 1992. Palaeolithic wooden artifacts from the Abric Romani (Capellades, Barcelona Spain). *Journal of Archaeological Science* 19:707–719.

Carlson, R. L. 1990. Cultural antecedents. In W. Suttles, ed., *Handbook of North American Indians*. Vol. 7: *Northwest Coast*, 60–69. Washington, D.C.: Smithsonian Institution.

———. 1996. Early Namu. In R. Carlson and L. Dalla Bona, eds., *Early human occupation in British Columbia*, 83–102. Vancouver: University of British Columbia Press.

Carter, L. D. 1981. A Pleistocene sand sea on the Alaskan Arctic Coastal Plain. *Science* 211:381–383.

Cassie, B. E., J. Brigham-Grette, N. W. Driscoll, M. S. Cook, and T. D. Herbert. 2004. Deglaciation to Holocene sea-ice history over Navarin and Pervenets canyons of the northern Bering Sea using diatoms and alkenones. Eos Trans. American Geophysical Union 85(47), Fall Meet. Suppl., abstract PP21B-1382.

Cattelain, P. 1997. Hunting during the Upper Paleolithic: Bow, spearthrower, or both? In H. Knecht, ed., *Projectile technology*, 213–240. New York: Plenum.

Catto, N. R. 1996. Richardson Mountains, Yukon-Northwest Territories: The northern portal of the postulated "ice-free corridor." *Quaternary International* 32:3–19.

CAVM Team. 2003. Circumpolar Arctic Vegetation Map. Scale 1:7,500,000. Conservation of Arctic Flora and Fauna (CAFF) Map No. 1. Anchorage: U.S. Fish and Wildlife Service.

Chard, C. S. 1974. *Northeast Asia in prehistory*. Madison: University of Wisconsin Press.

Chase, P. G. 1986. *The hunters of Combe Grenal: Approaches to middle Paleolithic subsistence in Europe*. International Series No. S-286. Oxford: British Archaeological Reports.

Chatters, J. C. 2000. The recovery and first analysis of an early Holocene human skeleton from Kennewick, Washington. *American Antiquity* 65(2):291–316.

Christensen, T., and J. Stafford. 1999. 1998 Cohoe Creek excavation project. *BC Association of Professional Consulting Archaeologists* 3(1):8.

———. 2005. Raised beach archaeology in northern Haida Gwaii: Preliminary results from the Cohoe Creek site. In Daryl W. Fedje and Rolf W. Mathewes, eds., *Haida Gwaii: Human history and environment from the time of the loon to the time of the iron people*, 245–273. Vancouver: University of British Columbia Press.

Churchill, S. E. 1998. Cold adaptation, heterochrony, and Neandertals. *Evolutionary Anthropology* 7(2):46–61.

Cinq-Mars, J. 1982. Les Grottes du Poisson-Bleu. *GEOS* 11(1):19–21.

———. 1990. La Place des Grottes du Poission-Bleu dans la prehistoire Beringienne. *Revista de Arquelogía Americana* 1:9–32.

Cinq-Mars, J., C. R. Harington, D. E. Nelson, and R. S. MacNeish. 1991. Engigstciak revisited: A note on early Holocene AMS dates from the "Buffalo Pit." *Canadian Archaeological Association Occasional Paper* 1:33–44.

Cinq-Mars, J., and R. E. Morlan. 1999. Bluefish Caves and Old Crow Basin: A new rapport. In R. Bonnichsen and K. L. Turnmire, eds., *Ice Age peoples of North America: Environments, origins, and adaptations of the first Americans*, 200–212. Corvallis: Oregon State University Press.

Clark, D. W. 1984. Northern fluted points: Paleo-Eskimo, Paleo-Arctic, or Paleo-Indian. *Canadian Journal of Anthropology* 4:65–81.

Clark, J. S., M. Lewis, J. S. McLachlan, and J. Hille Ris-Lambers. 2003. Estimating population spread: What can we forecast and how well? *Ecology* 84:1979–1988.

Clark, P. U., A. M. McCabe, A. C. Mix, and A. J. Weaver. 2003. Rapid rise of sea level 19,000 years ago and its global implications. *Science* 304:1141–1144.

Clark, P. U., and A. C. Mix. 2002. Ice sheets and sea level of the Last Glacial Maximum. *Quaternary Science Reviews* 21:1–7.

Colinvaux, P. A. 1967. Quaternary vegetational history of Arctic Alaska. In D. M. Hopkins, ed., *The Bering Land Bridge*, 207–231. Stanford, Calif.: Stanford University Press.

———. 1981. Historical ecology in Beringia: The south land bridge coast at St. Paul Island. *Quaternary Research* 16:18–36.

Colinvaux, P. A., and F. H. West. 1984. The Beringian ecosystem. *Quarterly Review of Archaeology* 5:10–16.

Cologne Radiocarbon Calibration Programme. 2005. CalPal-SFCP-2005 glacial calibration curve. Available at http://www.calpal.de.

Cook, J. P. 1969. The early prehistory of Healy Lake. Ph.D. diss., University of Wisconsin.

———. 1989. Historic archaeology and ethnohistory at Healy Lake, Alaska. *Arctic* 42(2):109–118.

———. 1996. Healy Lake. In F. W. West, ed., *American beginnings: The prehistory and paleoecology of Beringia*, 323–327. Chicago: University of Chicago Press.

Coon, C. S. 1962. *The origin of races*. New York: Alfred A. Knopf.

Coope, G. R., G. Lemdahl, J. J. Lowe, and A. Walkling. 1998. Temperature gradients in northern Europe during the last glacial-Holocene transition (14–9 ^{14}C kyr BP) interpreted from coleopteran assemblages. *Journal of Quaternary Science* 13:419–433.

Creager, J. S., and D. A. McManus. 1967. Geology of the floor of Bering and Chukchi Seas: American studies. In D. M. Hopkins, ed., *The Bering Land Bridge*, 7–31. Stanford, Calif.: Stanford University Press.

Croes, D. R. 1995. *The Hoko River archeological site complex: The wet/dry site (45 CA213), 3,000–1,700 BP*. Pullman: Washington State University Press.

Dall, W. H., and G. D. Harris. 1892. *Correlation papers, Neogene*. Bulletin No. 84. Washington, D.C.: U.S. Geological Survey.

Daly, R. A. 1934. *The changing world of the Ice Age*. New Haven, Conn.: Yale University Press.

Dansgaard, W., S. J. Johnson, H. B. Clausen, D. Dahl-Jensen, N. S. Gundestrup, C. U. Hammer, C. S. Hvidberg, J. P. Steffensen, A. E. Sveinbjornsdottir, J. Jouzel, and G. Bond. 1993. Evidence for general instability of past climate from a 250-kyr ice-core record. *Nature* 364:218–220.

Darwin, C. 1859. *The origin of species*. London: John Murray.

Davidson, I., and W. Noble. 1989. The archaeology of perception: Traces of depiction and language. *Current Anthropology* 30(2):125–155.

Dawson, G. M. 1894. Geologic notes on some of the coasts and islands of Bering Sea and vicinity. *Geological Society of America Bulletin* 5:117–146.

de Beaune, S., and R. White. 1993. Ice Age lamps. *Scientific American* 268: 108–113.

Dennell, R., and W. Roebroeks. 2005. An Asian perspective on early human dispersal from Africa. *Nature* 438:1099–1104.

Derevianko, A. P. et al. 1998. *Arkheologiya, geologiya i paleogeografiya pleistotsena i golotsena Gornogo Altaya*. Novosibirsk: Institute of Archaeology and Ethnography, Siberian Division, Russian Academy of Sciences.

d'Errico, F. 2003. The invisible frontier: A multiple species model for the origin of behavioral modernity. *Evolutionary Anthropology* 12:188–202.

d'Errico, F., C. Henshilwood, G. Lawson, M. Vanhaeren, A-M. Tillier, M. Soressi, F. Bresson, B. Maureille, A. Nowell, J. Lakarra, L. Backwell, and M. Julien. 2003. Archaeological evidence for the emergence of language, symbolism, and music: An alternative multidisciplinary perspective. *Journal of World Prehistory* 17:1–70.

Dikov, N. N. 1967. Otkrytie paleolita na Kamchatke i problema pervonachal'nogo zaseleniya Ameriki. In *Istoriya i kul'tura narodov Severa Dal'nego Vostoka*, 16–31. Moscow: Nauka.

———. 1969. Verkhnii paleolit Kamchatki. *Sovetskaya arkheologiya* 3:93–109.

———. 1977. *Arkheologicheskie pamyatniki Kamchatki, Chukotki i Verkhnei Kolymy.* Moscow: Nauka.

———. 1979. *Drevnie kul tury Severo-Vostochnoi Azii.* Moscow: Nauka.

———. 1990. Population migration from Asia to Pre-Columbian America. Paper presented at the 17th International Congress of Historical Sciences, Madrid, Spain.

———. 1996. The Ushki sites, Kamchatka Peninsula. In F. H. West, ed., *American beginnings: The prehistory and paleoecology of Beringia,* 244–250. Chicago: University of Chicago Press.

Dikov, N. N., and E. E. Titov. 1984. Problems of the stratification and periodization of the Ushki Sites. *Arctic Anthropology* 21(2):69–80.

Dillehay, T. D. 1989. Monte Verde. *Science* 245:1436.

Dixon, E. J. 1975. The Gallagher Flint Station, an Early Man site on the North Slope, Arctic Alaska and its role in relation to the Bering Land Bridge. *Arctic Anthropology* 12(1):68–75.

———. 1999. *Bones, boats and bison: Archeology and the first colonization of western North America.* Albuquerque: University of New Mexico Press.

———. 2001. Human colonization of the Americas: Timing, technology and process. *Quaternary Science Reviews* 20:277–299.

Dixon, E. J., T. H. Heaton, T. E. Fifield, T. D. Hamilton, D. E. Putnam, and F. Grady. 1997. Late Quaternary regional geoarchaeology of Southeast Alaska karst: A progress report. *Geoarchaeology* 12(6):689–712.

Dixon, E. J., and G. S. Smith. 1986. Broken canines from Alaskan cave deposits: Re-evaluating evidence for domesticated dog and early humans in Alaska. *American Antiquity* 51:341–351.

Dolukhanov, P., D. Sokoloff, and A. Shukurov. 2001. Radiocarbon chronology of Upper Palaeolithic sites in Eastern Europe at improved resolution. *Journal of Archaeological Science* 28:699–712.

Dumond, D. E. 1980. The archaeology of Alaska and the peopling of America. *Science* 209:984–991.

———. 2001. The archaeology of eastern Beringia: Some contrasts and connections. *Arctic Anthropology* 38(2):196–205.

Dyke, A. S. 2004. An outline of North American deglaciation with emphasis on central and northern Canada. In J. Ehlers and P. L. Gibbard, eds., *Quaternary glaciation: Extent and Chronology. Part II: North America,* 373–424. Amsterdam: Elsevier.

Easton, N. A. 1992. Mal de mer above terra incognita, or "what ails the coastal migration theory?" *Arctic Anthropology* 29(2):28–41.

Easton, N. A., G. MacKay, C. Baker, K. Hermanson, and P. Schnurr. 2004. Nenana in Canada: 2003 excavations of a multi-component site (KdVo-6) in the Mirror Creek Valley, Yukon Territory, Canada. Paper presented at the 31st annual meeting of the Alaska Anthropological Association, Whitehorse, Yukon, 7–10 April.

Edwards, M. E., L. B. Brubaker, A. V. Lozhkin, and P. M. Anderson. 2005. Structurally novel biomes: A response to past warming in Beringia. *Ecology* 86: 1696–1703.

Edwards, M. E., J. C. Dawe, and W. S. Armbruster. 1991. Pollen size of Betula in northern Alaska and the interpretation of late-Quaternary pollen records. *Canadian Journal of Botany* 69:1666–1672.

Edwards, M. E., C. J. Mock, B. P. Finney, V. A. Barber, and P. J. Bartlein. 2001. Potential analogues for paleoclimatic variations in eastern interior Alaska during the past 14,000 years: Atmospheric-circulation controls of regional temperature and moisture responses. *Quaternary Science Reviews* 20: 189–202.

Elias, S. A. 1992. Late Wisconsin insects and plant macrofossils associated with the Colorado Creek mammoth, southwestern Alaska: Taphonomic and paleoenvironmental implications. *22nd Arctic Workshop Program and Abstracts* (University of Colorado), 45–47.

———. 2000. Late Pleistocene climates of Beringia, based on fossil beetle analysis. *Quaternary Research* 53:229–235.

———. 2001a. Mutual climatic range reconstructions of seasonal temperatures based on late Pleistocene fossil beetle assemblages in eastern Beringia. *Quaternary Science Reviews* 20:77–91.

———. 2001b. Beringian paleoecology: Results from the 1997 workshop. *Quaternary Science Reviews* 20:7–13.

Elias, S. A., K. H. Anderson, and J. T. Andrews. 1996. Late Wisconsin climate in northeastern USA and southeastern Canada, reconstructed from fossil beetle assemblages. *Journal of Quaternary Science* 11:417–421.

Elias, S. A., S. K. Short, and H. H. Birks. 1997. Late Wisconsin environments of the Bering Land Bridge. *Palaeogeography, Palaeoclimatology, Palaeoecology* 136:293–308.

Elias, S. A., S. K. Short, C. H. Nelson, and H. H. Birks. 1996. Life and times of the Bering Land Bridge. *Nature* 382:60–63.

Elias, S. A., S. K. Short, and R. L. Phillips. 1992. Paleoecology of Late-Glacial peats from the Bering Land Bridge, Chukchi Sea Shelf, Northwestern Alaska. *Quaternary Research* 38:371–378.

Enard, W., M. Przeworski, S. E. Fisher, C. S. L. Lai, V. Wiebe, T. Kitano, A. P. Monaco, and S. Paabo. 2002. Molecular evolution of FOXP2, a gene involved in speech and language. *Nature* 418:869–872.

Engstrom, D. R., B. C. S. Hansen, and H. E. Wright. 1990. A possible Younger Dryas record in southeastern Alaska. *Science* 250:1383–1385.

Erlandson, J., R. Walser, H. Maxwell, N. Bigelow, J. Cook, R. Lively, C. Adkins, D. Dodson, A. Higgs, and J. Wilber. 1991. Two early sites of eastern Beringia: Context and chronology in Alaskan interior archaeology. *Radiocarbon* 33(1):46–53.

Ermolova, N. M. 1978. *Teriofauna doliny Angary v pozdnem antropogene.* Novosibirsk: Nauka.

Fagan, B. M. 1990. *The journey from Eden: The peopling of our world.* London: Thames and Hudson.

———. 2004. *The great journey: The peopling of Ancient America.* 2nd ed. Gainesville: University Press of Florida.

Fairbanks, R. G. 1989. A 17,000-year glacio-eustatic sea level record: Influence of glacial melting rates on the Younger Dryas event and deep-ocean circulation. *Nature* 342:637–642.

Ferguson, D. E. 1997. Gallagher Flint Station Locality 1: A reappraisal of a proposed late Pleistocene site on the Sagavanirktok River Valley, Arctic Alaska. M.A. thesis, Department of Anthropology, University of Alaska, Fairbanks.

Fiedel, S. J. 1999. Older than we thought: Implications of corrected dates for Paleoindians. *American Antiquity* 64(1):95–115.

———. 2002. Initial human colonization of the Americas: An overview of the issues and the evidence. *Radiocarbon* 44(2):407–436.

Fladmark, K. R. 1978. The feasibility of the northwest coast as a migration route for Early Man. In A. L. Bryan, ed., *Early Man in America from a circum-Pacific perspective*, 119–128. Occasional Paper No. 1. Alberta: Department of Anthropology, University of Alberta.

———. 1979. Routes: Alternative migration corridors for Early Man in North America. *American Antiquity* 44(1):55–69.

———. 1983. Time and places: Environmental correlates of mid-to-late Wisconsinan human population expansion in North America." In R. Shutler, ed., *Early Man in the New World*, 13–41. Beverly Hills, Calif.: Sage.

Fladmark, K. R., J. C. Driver, and D. Alexander. 1988. The Paleoindian component at Charlie Lake Cave (HbRf 39), British Columbia. *American Antiquity* 53(2):371–384.

Flenniken, J. J. 1987. The Paleolithic Dyuktai pressure blade technique of Siberia. *Arctic Anthropology* 24(2):117–132.

Forde, C. D. 1934. *Habitat, economy and society: A geographical introduction to ethnology.* London: Methuen.

Gabunia, L. et al. 2000. Earliest Pleistocene cranial remains from Dmanisi, Republic of Georgia: Taxonomy, geological setting, and age. *Science* 288:1019–1025.

Gallant, A. L., E. F. Binnian, J. M. Omernik, and M. B. Shasby. 1995. *Ecoregions of Alaska.* Professional Paper No. 1567. Washington, D.C.: U.S. Geological Survey.

Gamble, C. 1986. *The Palaeolithic settlement of Europe.* Cambridge: University of Cambridge Press.

———. 1994. *Timewalkers: The prehistory of global colonization.* Cambridge: Harvard University Press.

———. 1995. The earliest occupation of Europe: The environmental background. In W. Roebroeks and T. van Kolfschoten, eds., *The earliest occupation of Europe,* 279–295. Leiden: University of Leiden.

Gening, V. F., and V. T. Petrin. 1985. *Pozdnepaleolitcheskaya epokha na Yuge Zapadnoi Sibiri.* Novosibirsk: Nauka.

Gerasimov, M. M. 1964. The Paleolithic site Malta: Excavations of 1956–1957. In H. N. Michael, ed., *The archaeology and geomorphology of Northern Asia: Selected works,* 3–32. Toronto: University of Toronto Press.

Gershanovich, D. E. 1967. Late Quaternary sediments of Bering Sea and the Gulf of Alaska. In D. M. Hopkins, ed., *The Bering land bridge,* 32–46. Stanford, Calif.: Stanford University Press.

Giterman, R. E., and L. V. Golubeva. 1967. Vegetation of eastern Siberia during the anthropogene period. In D. M. Hopkins, ed., *The Bering land bridge,* 232–244. Stanford, Calif.: Stanford University Press.

Glushkova, O. Y. 2001. Geomorphical correlation of late Pleistocene glacial complexes of western and eastern Beringia. *Quaternary Science Reviews* 20:405–417.

Goebel, T. 1999. The Pleistocene colonization of Siberia and peopling of the Americas: An ecological approach. *Evolutionary Anthropology* 8:208–227.

———. 2001a. Siberian early Upper Paleolithic. In P. N. Peregrine and M. Ember, eds., *Encyclopedia of prehistory.* Vol. 2: *Arctic and Subarctic,* 181–185. New York: Kluwer Academic/Plenum.

———. 2001b. Siberian middle Upper Paleolithic. In P. N. Peregrine and M. Ember, eds., *Encyclopedia of prehistory.* Vol. 2: *Arctic and Subarctic,* 192–196. New York: Kluwer Academic/Plenum.

———. 2002. The "microblade adaptation" and recolonization of Siberia during the late Upper Pleistocene. In R. G. Elston and S. L. Kuhn, eds., *Thinking small: Global perspectives on microlithization,* 117–131. Archaeological Paper No. 12. Washington, D.C.: American Anthropological Association.

Goebel, T., and N. Bigelow. 1996. Panguingue Creek. In F. H. West, ed., *American beginnings: The prehistory and paleoecology of Beringia,* 366–371. Chicago: University of Chicago Press.

Goebel, T., A. P. Derevianko, and V. T. Petrin. 1993. Dating the middle-to-upper-Paleolithic transition at Kara-Bom. *Current Anthropology* 34:452–458.

Goebel, T., and R. Powers. 1989. A possible Paleoindian dwelling in the Nenana Valley, Alaska: Spatial analysis at the Walker Road site. Paper presented at the 16th annual meeting of the Alaska Anthropological Association, Anchorage, Alaska, 3–4 March.

Goebel, T., W. R. Powers, and N. H. Bigelow. 1991. The Nenana complex of Alaska and Clovis origins. In R. Bonnichsen and K. L. Turnmire, eds., *Clovis: Origins and adaptations*, 49–79. Corvallis, Ore.: Center for the Study of the First Americans.

Goebel, T., W. R. Powers, N. H. Bigelow, and A. S. Higgs. 1996. Walker Road. In F. H. West, ed., *American beginnings: The prehistory and paleoecology of Beringia*, 356–363. Chicago: University of Chicago Press.

Goebel, T., and S. Slobodin. 1999. The colonization of western Beringia: Technology, ecology, and adaptation. In R. Bonnichsen and K. L. Turnmire, eds., *Ice Age peoples of North America: Environments, origins, and adaptations of the first Americans*, 104–155. Corvallis: Oregon State University Press.

Goebel, T., M. R. Waters, I. Buvit, M. V. Konstantinov, and A. V. Konstantinov. 2000. Studenoe-2 and the origins of microblade technologies in the Transbaikal, Siberia. *Antiquity* 74:567–575.

Goebel, T., M. R. Waters, and M. Dikova. 2003. The archaeology of Ushki Lake, Kamchatka, and the Pleistocene peopling of the Americas. *Science* 301:501–505.

Goebel, T., M. R. Waters, and M. N. Meshcherin. 2001. Masterov Kliuch and the early Upper Paleolithic of the Transbaikal, Siberia. *Asian Perspectives* 39(1–2):47–70.

Goetcheus, V., and H. Birks. 2001. Full-glacial upland tundra vegetation preserved under tephra in the Beringia National Park, Seward Peninsula, Alaska. *Quaternary Science Reviews* 20:135–147.

Goldberg, P., S. Weiner, O. Bar-Yosef, Q. Xu, and J. Liu. 2001. Site formation processes at Zhoukoudian, China. *Journal of Human Evolution* 41:483–530.

Goren-Inbar, N., N. Alperson, M. E. Kislev, O. Simchoni, Y. Melamed, A. Ben-Nun, and E. Werker. 2004. Evidence of hominin control of fire at Gesher Benot Ya'aqov, Israel. *Science* 304:725–727.

Grayson, D. K., and D. J. Meltzer. 2003. A requiem for North American overkill. *Journal of Archaeological Science* 30:585–593.

Greenberg, J. H., C. G. Turner, and S. L. Zegura. 1986. The settlement of the Americas: A comparison of the linguistic, dental and genetic evidence. *Current Anthropology* 27:208–227.

Grichuk, V. P. 1984. Late Pleistocene vegetation history. In A. A. Velichko, ed., *Late Quaternary environments of the Soviet Union*, 155–178. Minneapolis: University of Minnesota Press.

Grosswald, M. G. 1988. An Antarctic-style ice sheet in the northern hemisphere: Toward a new global glacial theory. *Polar Geology and Geography* 12:239–267.

Gruhn, R. 1994. The Pacific Coast route of initial entry: An overview. In R. Bonnichsen and D. G. Steele, eds., *Methods and theory for investigating the peopling of the Americas*, 249–256. Corvallis, Ore.: Center for the Study of the First Americans.

Gubser, N. 1965. *The Nunamiut Eskimo: Hunters of caribou.* New Haven, Conn.: Yale University Press.

Guthrie, R. D. 1968. Paleoecology of the large-mammal community in interior Alaska during the late Pleistocene. *American Midland Naturalist* 79:346–463.

———. 1982. Mammals of the mammoth steppe as paleoenvironmental indicators. In D. M. Hopkins, J. V. Matthews Jr., C. E. Schweger, and S. B. Young, eds., *Paleoecology of Beringia*, 307–326. New York: Academic Press.

———. 1983a. Paleoecology of the site and its implication for early hunters. In W. R. Powers, R. D. Guthrie, and J. F. Hoffecker, eds., *Dry Creek: The archaeology and paleoecology of a late Pleistocene Alaskan hunting camp*, 209–287. Anchorage: U.S. National Park Service.

———. 1983b. Appendix B: Composite bone-stone tool reproduction and testing. In W. R. Powers, R. D. Guthrie and J. F. Hoffecker, eds., *Dry Creek: The archeology and paleoecology of a late Pleistocene Alaskan hunting camp*, 348–374. Anchorage: U.S. National Park Service.

———. 1990. *Frozen fauna of the mammoth steppe: The story of Blue Babe.* Chicago: University of Chicago Press.

———. 2001. Origin and causes of the mammoth steppe: A story of cloud cover, woolly mammoth tooth pits, buckles, and inside-out Beringia. *Quaternary Science Reviews* 20:549–574.

———. 2003. Rapid body size decline in Alaskan Pleistocene horses before extinction. *Nature* 426:169–171.

———. 2004. Radiocarbon evidence of mid-Holocene mammoths stranded on an Alaskan Bering Sea island. *Nature* 429:746–749.

———. 2006. New carbon dates link climatic change with human colonization and Pleistocene extinctions. *Nature* 441:207–209.

Hadingham, E. 2004. America's first immigrants. *Smithsonian* 35(8):90–98.

Hamilton, T. D. 1994. Late Cenozoic glaciation of Alaska. In G. Plafker and H. Berg, eds., *The geology of North America*. Vol. G-1: *The geology of Alaska*, 813–844. Boulder, Colo.: Geological Society of America.

Hamilton, T. D., G. M. Ashley, K. M. Reed, and C. E. Schweger. 1993. Late Pleistocene vertebrates and other fossils from Epiguruk, northwestern Alaska. *Quaternary Research* 39:381–389.

Hamilton, T. D., and T. Goebel. 1999. Late Pleistocene peopling of Alaska. In R. Bonnichsen and K. L. Turnmire, eds., *Ice Age peoples of North America: Environments, origins, and adaptations of the first Americans*, 156–199. Corvallis: Oregon State University Press.

Hamilton, T. D., and S. C. Porter. 1975. Itkillik glaciation in the Brooks Range, northern Alaska. *Quaternary Research* 5:471–498.

Hansen, B. C. S., and D. R. Engstrom. 1996. Vegetation history of Pleasant Island, southeastern Alaska, since 13,000 yr B.P. *Quaternary Research* 46:161–175.

Hare, P. G. , S. Greer, R. Gotthardt, R. Farnell, V. Bowyer, and C. Schweger. 2004. Multidisciplinary investigations of alpine ice patches in southwest Yukon, Canada: Ethnographic and archaeological investigations. *Arctic* 57(3):260–272.

Hatt, G. 1969. Arctic skin clothing in Eurasia and America: An ethnographic study. *Arctic Anthropology* 5(2):3–132.

Hauser, M. D., N. Chomsky, and W. T. Fitch. 2002. The faculty of language: What is it, who has it, and how did it evolve? *Science* 298:1569–1579

Haynes, C. V. 1982. Were Clovis progenitors in Beringia? In D. M. Hopkins, J. V. Matthews Jr., C. E. Schweger, and S. B. Young, eds., *Paleoecology of Beringia*, 383–398. New York: Academic Press.

Haynes, G. 1983a. Frequencies of spiral and green-bone fractures on ungulate limb bones in modern surface assemblages. *American Antiquity* 48(1):102–114.

———. 1983b. A guide for differentiating mammalian carnivore taxa responsible for gnaw damage to herbivore limb bones. *Paleobiology* 9(2):164–172.

Heilprin, A. 1887. *The geographical and geological distribution of animals.* New York: Appleton.

Heiser, P. A., and J. J. Roush. 2001. Pleistocene glaciations in Chukotka, Russia: Moraine mapping using satellite synthetic aperture radar (SAR) imagery. *Quaternary Science Reviews* 20:393–404.

Henderson-Sellers, A., and P. J. Robinson. 1986. *Contemporary climatology.* Edinburgh Gate: Addison Wesley Longman.

Hey, J. 2005. On the number of New World founders: A population genetic portrait of the peopling of the Americas. *PLoS (Public Library of Science) Biology* 3(6), file e193. Available at http://biology.plosjournals. org/perlserv/?request=get-document & doi=10.1371/journal.pbio.0030193 (accessed September 15, 2006).

Hoffecker, J. F. 1983. A description and analysis of artifact clusters in Components I and II at the Dry Creek site. In W. R. Powers, R. D. Guthrie, and J. F. Hoffecker, eds., *Dry Creek: The archeology and paleoecology of a late Pleistocene Alaskan hunting camp*, 307–347. Anchorage: U.S. National Park Service.

———. 1988. Applied geomorphology and archaeological survey strategy for sites of Pleistocene age: An example from Central Alaska. *Journal of Archaeological Science* 15:683–713.

———. 1996. Moose Creek. In F. H. West, ed., *American beginnings: The prehistory and paleoecology of Beringia*, 363–366. Chicago: University of Chicago Press.

———. 2001. Late Pleistocene and early Holocene sites in the Nenana River Valley, central Alaska. *Arctic Anthropology* 38(2):139–153.

———. 2002a. *Desolate landscapes: Ice-Age settlement of eastern Europe.* New Brunswick, N.J.: Rutgers University Press.

———. 2002b. The eastern Gravettian "Kostenki Culture" as an arctic adaptation. *Anthropological Papers of the University of Alaska, NS* 2(1):115–136.

———. 2005a. *A prehistory of the north: Human settlement of the higher latitudes.* New Brunswick, N.J.: Rutgers University Press.

———. 2005b. Innovation and technological knowledge in the Upper Paleolithic of northern Eurasia. *Evolutionary Anthropology* 14:186–198.

Hoffecker, J. F., and N. Cleghorn. 2000. Mousterian hunting patterns in the northwestern Caucasus and the ecology of the Neanderthals. *International Journal of Osteoarchaeology* 10:368–378.

Hoffecker, J. F., and S. A. Elias. 2003. Environment and archeology in Beringia. *Evolutionary Anthropology* 12(1):34–49.

———. 2005. Woody shrubs may have been critical to the Lateglacial settlement of Beringia. Paper presented at the 32nd annual meeting of the Alaska Anthropological Association, Anchorage, Alaska, 10–12 March.

Hoffecker, J. F., W. R. Powers, and N. H. Bigelow. 1996. Dry Creek. In F. H. West, ed., *American beginnings: The prehistory and paleoecology of Beringia,* 343–352. Chicago: University of Chicago Press.

Hoffecker, J. F., W. R. Powers, and T. Goebel. 1993. The colonization of Beringia and the peopling of the New World. *Science* 259:46–53.

Hoffecker, J. F., W. R. Powers, and P. G. Phippen. 1996. Owl Ridge. In F. H. West, ed., *American beginnings: The prehistory and paleoecology of Beringia,* 353–356. Chicago: University of Chicago Press.

Höfle, C., and C.-L. Ping. 1996. Properties and soil development of late-Pleistocene paleosols from Seward Peninsula, northwest Alaska. *Geoderma* 71:219–243.

Holliday, T. W. 1999. Brachial and crural indices of European late Upper Paleolithic and Mesolithic humans. *Journal of Human Evolution* 36:549–566.

Holliday, V. T. 2000. The evolution of Paleoindian geochronology and typology on the Great Plains. *Geoarchaeology* 15(3):227–290.

Holloway, R. L. 1985. The poor brain of *Homo sapiens neanderthalensis:* See what you please … In E. Delson, ed., *Ancestors: The hard evidence,* 319–324. New York: Alan R. Liss.

Holmes, C. E. 1974. New evidence of a late Pleistocene culture in central Alaska: Preliminary investigations at Dry Creek. Paper presented at the 7th annual meeting of the Canadian Archaeological Association, Whitehorse, Canada.

———. 1996. Broken Mammoth. In F. H. West, ed., *American beginnings: The prehistory and paleoecology of Beringia,* 312–318. Chicago: University of Chicago Press.

———. 1998. New data pertaining to Swan Point, the oldest microblade site known in Alaska. *Current Research in the Pleistocene* 15:21–22.

———. 2001. Tanana River Valley archaeology circa 14,000 to 9000 B.P. *Arctic Anthropology* 38(2):154–170.

Holmes, C. E., and B. A. Crass. 2003. Early cultural components in central Alaska: An update from Swan Point. Paper presented at the 30th annual meeting of the Alaska Anthropological Association, Fairbanks, Alaska, 26–29 March.

Holmes, C. E., R. VanderHoek, and T. E. Dilley. 1996. Swan Point. In F. H. West, ed., *American beginnings: The prehistory and paleoecology of Beringia*, 319–322. Chicago: University of Chicago Press.

Hopkins, D. M. 1959. Cenozoic history of the Bering Land Bridge. *Science* 129:1519–1528.

———. 1967a. Introduction. In D. M. Hopkins, ed., *The Bering land bridge*, 1–6. Stanford, Calif.: Stanford University Press.

———. 1967b. Preface. In D. M. Hopkins, ed., *The Bering land bridge*, vii–ix. Stanford, Calif.: Stanford University Press.

———. 1967c. The Cenozoic history of Beringia: A synthesis. In D. M. Hopkins, ed., *The Bering land bridge*, 451–484. Stanford, Calif.: Stanford University Press.

———. 1973. Sea level history of Beringia during the past 250 000 years. *Quaternary Research* 3:520–540.

———. 1982. Aspects of the paleogeography of Beringia during the late Pleistocene. In D. M. Hopkins, J. V. Matthews Jr., C. E. Schweger, and S. B. Young, eds., *Paleoecology of Beringia*, 3–28. New York: Academic Press.

Hopkins, D. M., J. V. Matthews Jr., C. E. Schweger, and S. B. Young, eds. 1982. *Paleoecology of Beringia*. New York: Academic Press.

Hopkins, D. M., P. A. Smith, and J. V. Matthews Jr. 1981. Dated wood from Alaska and the Yukon: Implications for forest refugia in Beringia. *Quaternary Research* 15:217–249.

Hu, F. S., L. B. Brubaker, and P. M. Anderson. 1995. Postglacial vegetation and climate change in the northern Bristol Bay region, southwestern Alaska. *Quaternary Research* 43:382–392.

Hublin, J.-J. 1998. Climatic changes, paleogeography, and the evolution of the Neandertals. In T. Akazawa, K. Aoki, and O. Bar-Yosef, eds., *Neandertals and modern humans in western Asia*, 295–310. New York: Plenum.

Hughen, K. A., J. R. Southon, S. J. Lehman, and J. T. Overpeck. 2000. Synchronous radiocarbon and climate shifts during the last deglaciation. *Science* 290:1951–1954.

Hughes, B. A., and T. J. Hughes. 1994. Transgressions: Rethinking Beringian glaciation. *Paleogeography, Paleoclimatology, Paleoecology* 110:275–294.

Hultén, E. 1937. *Outline of the history of arctic and boreal biota during the Quaternary period*. New York: Lehre J. Cramer.

———. 1958. *The amphi-atlantic plants and their phytogeographical connections*. Series 7. Stockholm: Almqvist and Wiksell.

———. 1968. *Flora of Alaska and neighboring territories*. Stanford, Calif.: Stanford University Press.

Irving, W. N. 1978. Pleistocene archaeology in eastern Beringia. In A. L. Bryan, ed., *Early Man in America from a circum-Pacific perspective*, 96–101. Occasional Paper No. 1. Alberta: Department of Anthropology, University of Alberta.

Irving, W. N., and C. R. Harington. 1973. Upper Pleistocene radiocarbon-dated artifacts from the northern Yukon. *Science* 179:335–340.

Irving, W. N., A. V. Jopling, and B. F. Beebe. 1986. Indications of Pre-Sangamon humans near Old Crow, Yukon, Canada. In A. L. Bryan, ed., *New evidence for the Pleistocene peopling of the Americas*, 49–63. Orono, Maine: Center for the Study of Early Man.

Jaccard, S. L., G. H. Haug, D. M. Sigman, T. F. Pedersen, H. R. Thierstein, and U. Röhl. 2005. Glacial/interglacial changes in subarctic North Pacific stratification. *Science* 308:1003–1006.

Jackson, L. E., and A. Duk-Rodkin. 1996. Quaternary geology of the ice-free corridor: Glacial controls on the peopling of the New World. In T. Akazawa and E. J. E. Szathmary, eds., *Prehistoric Mongoloid dispersals*, 214–227. New York: Oxford University Press.

Johnsen, S. J. et al. 1997. The delta O-18 record along the Greenland Ice Core Project deep ice core and the problem of possible Eemian climatic instability. *Journal of Geophysical Research* 102:26397–26410.

Jordan, J. W. 2001. Late Quaternary sea level change in southern Beringia: Postglacial emergence of the western Alaska Peninsula. *Quaternary Science Reviews* 20:509–523.

Josenhans, H., D. Fedje, R. Pienitz, and J. Southon. 1997. Early humans and rapidly changing Holocene sea levels in the Queen Charlotte Islands–Hecate Strait, British Columbia, Canada. *Science* 277:71–74.

Kelly, Robert L. 1995. *The foraging spectrum: Diversity in hunter-gatherer lifeways*. Washington, D.C.: Smithsonian Institution.

Kind, N. V. 1974. *Geokhronologiya pozdnego antropologena po izotopnym dannym*. Moscow: Nauka.

King, Maureen L., and Sergei B. Slobodin. 1994. Terminal Pleistocene occupation of the Kheta site, Upper Kolyma region, northeastern Russia. *Current Research in the Pleistocene* 11:138–140.

———. 1996. A fluted point from the Uptar site, northeastern Siberia. *Science* 273:634–636.

Kinyamario, J. I., and S. K. Imbamba. 1992. Savanna at Nairobi National Park, Nairobi. In S. P. Long, M. B. Jones, and M. J. Roberts, eds., *Primary productivity of grass ecosystems of the tropics and sub-tropics*, 25–69. London: Chapman and Hall.

Kir'yak, M. A. 1993. *Arkheologiya Zapadnoi Chukotki*. Moscow: Nauka.

———. 1996. Bol'shoi Elgakhchan 1 and 2, Omolon River Basin, Magadan District. In F. H. West, ed., *American beginnings: The prehistory and paleoecology of Beringia*, 228–236. Chicago: University of Chicago Press.

Kitagawa, H., and J. van der Plicht. 1998. Atmospheric radiocarbon calibration to 45,000 yr B.P.: Late Glacial fluctuations and cosmogenic isotope production. *Science* 279:1187–1190.

Klein, R. G. 1973. *Ice-Age hunters of the Ukraine.* Chicago: University of Chicago Press.

———. 1999. *The human career.* 2nd ed. Chicago: University of Chicago Press.

Knopf, A. 1910. The probably Tertiary land connection between Asia and North America. *University of California Publications in Geology Bulletin* 5:413–420.

Kojima, S., and R. C. Brooke. 1986. *An annotated vascular flora of areas adjacent to the Dempster Highway, Central Yukon Territory.* Contributions to Natural Science No. 3. Victoria: British Columbia Provincial Museum.

Kolesnikov, B. P. 1955. Zonal'nost' rastitel'nogo pokrova na Dal'nem Vostoke i problemy vosstonavleniya i zashchity lesnogo poyasa. In *Problemy razvitiya lesnoi i derevoobrabatyvayushchei promyshlennosti na Dal'nem Vostoke.* Moscow: Akademiya nauk SSSR.

Konyukova, L. G., V. V. Orlova, and Ts. A. Shver. 1971. *Klimaticheskie kharakteristiki SSSR po mesyatsam.* Leningrad: Gidrometeoizdat.

Krestov, P. V. 2003. Forest vegetation of easternmost Russia (Russian Far East). In J. Kolbek, M. Srutek, and E. Box, eds., *Forest vegetation of Northeast Asia,* 93–180. Dordrecht: Kluwer Academic.

Kunz, M. L., M. Bever, and C. Adkins. 2003. *The Mesa site: Paleoindians above the Arctic Circle.* Open File Report No. 86. Anchorage: Bureau of Land Management Alaska.

Kunz, M. L., and R. E. Reanier. 1994. Paleoindians in Beringia: Evidence from arctic Alaska. *Science* 263:660–662.

———. 1996. The Mesa site, Iteriak Creek. In F. W. West, ed., *American beginnings: The prehistory and paleoecology of Beringia,* 497–504. Chicago: University of Chicago Press.

Kuzmin, Y. V., and S. G. Keates. 2005. Dates are not data: Paleolithic settlement patterns in Siberia derived from radiocarbon records. *American Antiquity* 70(4):773–789.

Lahr, M. M., and R. Foley. 1994. Multiple dispersals and modern human origins. *Evolutionary Anthropology* 3:48–60.

Larsen, H. 1968. Trail Creek: Final report of the excavation of two caves on Seward Peninsula, Alaska. *Acta Arctica* 15:7–79.

Laughlin, W. S. 1967. Human migration and permanent occupation in the Bering Sea area. In D. M. Hopkins, ed., *The Bering land bridge,* 409–450. Stanford, Calif.: Stanford University Press.

———. 1975. Aleuts: Ecosystem, Holocene history, and Siberian origin. *Science* 189:507–515.

Lee, C. M., and E. J. Dixon. 2006. Assessing the origin and function of early period microblade technology in southeast Alaska. Program and Abstracts, 99–100, 36th International Arctic Workshop, Institute of Arctic and Alpine Research (INSTAAR), University of Colorado at Boulder.

Lieth, H. 1975. Primary production of the major vegetation units of the world. In H. Lieth and R. H. Whittaker, eds., *Primary productivity of the biosphere*, 203–215. New York: Springer-Verlag.

Lisiecki, L. E., and M. E. Raymo. 2005. A Pliocene-Pleistocene stack of 57 globally distributed benthic δ18O records. *Paleoceanography* 20: PA1003 doi:10.1029/2004PA001071.

Lively, R. A. 1988. A study of the effectiveness of a small scale probabilistic sampling design at an interior Alaska site, Chugwater (FAI-035). Report on file, U.S. Army Corps of Engineers, Alaska District, Anchorage.

———. 1996. Chugwater. In F. H. West, ed., *American beginnings: The prehistory and paleoecology of Beringia*, 308–311. Chicago: University of Chicago Press.

Lozhkin, A. V., P. M. Anderson, B. V. Belaya, and T. V. Stetsenko. 2002. Reflections on modern pollen rain of Chukotka from bottom lake sediments. In K. V. Simakov, ed., *Quaternary palaeogeography of Beringia*, 40–50. Magadan, Russia: NEISRI FEB RAS.

Lozhkin, A. V., P. M. Anderson, W. R. Eisner, L.G. Ravako, D. M. Hopkins, L.B. Brubaker, P. A. Colinvaux, and M. C. Miller. 1993. Late Quaternary lacustrine pollen records from southwestern Beringia. *Quaternary Research* 39:314–324.

Lozhkin, A. V., P. M. Anderson, S.L. Vartanyan, T.A. Brown, B.V. Belaya, and A. N. Kotov. 2001. Late Quaternary paleoenvironments and modern pollen data from Wrangel Island (northern Chukotka). *Quaternary Science Reviews* 20:217–233.

Lydolph, P. 1977. *Geography of the U.S.S.R.* 3rd ed. New York: Wiley.

Mackay, J. R., W. H. Mathews, and R. S. MacNeish. 1961. Geology of the Engigstciak archaeological site, Yukon Territory. *Arctic* 14(1):25–52.

MacNeish, R. S. 1954. The Pointed Mountain site near Fort Laird, Northwest Territories, Canada. *American Antiquity* 19(3):234–253.

———. 1956. The Engigstciak site on the Yukon arctic coast. *Anthropological Papers of the University of Alaska* 4(2):91–111.

———. 1963. The early peopling of the New World. *Anthropological Papers of the University of Alaska* 10(2):93–134.

Maddren, A. G. 1905. *Smithsonian exploration in Alaska in 1904, in search of mammoth and other fossil remains*. Miscellaneous Collections No. 44. Washington, D.C.: Smithsonian Institution.

Maitland, R. E. 1986. The Chugwater site (FAI-035), Moose Creek Bluff, Alaska. Final Report, 1982 and 1983 seasons. Report on file, U.S. Army Corps of Engineers, Alaska District, Anchorage.

Malhi, R. S., J. A. Eshleman, J. A. Greenberg, D. A. Weiss, B. A. Schultz Shook, F. A. Kaestle, J. G. Lorenz, B. M. Kemp, J. R. Johnson, and D. G. Smith. 2002. The structure of diversity within New World mitochondrial DNA haplogroups: Implications for the prehistory of North America. *American Journal of Human Genetics* 70:905–919.

Mandryk, C. A. S., H. Josenhans, D. W. Fedje, and R. W. Mathewes. 2001. Late Quaternary paleoenvironments of northwestern North America: Implications for inland versus coastal migration routes. *Quaternary Science Reviews* 20: 301–314.

Manley, W. F. 2002. *Postglacial flooding of the Bering Land Bridge: A geospatial animation.* INSTAAR, University of Colorado, VI, http://instaar.colorado.edu/QGISL/bering_land_bridge (accessed September 14, 2006).

Mann, D. H., and T. D. Hamilton. 1995. Late Pleistocene and Holocene paleoenvironments of the north Pacific coast. *Quaternary Science Reviews* 14:449–471.

Mann, D. H., and D. M. Peteet. 1994. Extent and timing of the Last Glacial Maximum in southwestern Alaska. *Quaternary Research* 42:136–148.

Mann, D. H., R. E. Reanier, D. M. Peteet, M. L. Kunz, and M. Johnson. 2001. Environmental change and Arctic Paleoindians. *Arctic Anthropology* 38(2): 119–138.

Marincovich, L., and A. Y. Gladenkov. 2001. New evidence for the age of Bering Strait. *Quaternary Science Reviews* 20:329–335.

Martin, P. S. 1989. Prehistoric overkill: The global model. In P. S. Martin and R. G. Klein, eds., *Quaternary extinctions*, 354–403. Tucson: University of Arizona Press.

Mason, O. K., P. M. Bowers, and D. M. Hopkins. 2001. The early Holocene Milankovitch thermal maximum and humans: Adverse conditions for the Denali complex in eastern Beringia. *Quaternary Science Reviews* 20:525–548.

Matheus, P., M. Kunz, and R. D. Guthrie. 2003. Using frequency distributions of radiocarbon dates to detect relative changes in Pleistocene mammal populations: A test case from northern Alaska. In *Third International Mammoth Conference, Program and Abstracts*, 3–5. Dawson, Canada: Yukon Territorial Government.

Matthews, J. V., Jr. 1968. A paleoenvironmental analysis of three late Pleistocene coleopterous assemblages from Fairbanks, Alaska. *Quaestiones Entomologicae* 4:202–224.

———. 1982. East Beringia during late Wisconsin time: A review of the biotic evidence. In D. M. Hopkins, J. V. Matthews Jr., C. E. Schweger, and S. B. Young, eds., *Paleoecology of Beringia*, 127–150. New York: Academic Press.

Matthews, J. V., Jr., and A. Telka. 1997. Insect fossils from the Yukon. In H. V. Danks and J. A. Downes, eds., *Insects of the Yukon*, 911–962. Ottawa: Biological Survey of Canada.

McBrearty, S., and A. S. Brooks. 2000. The revolution that wasn't: A new interpretation of the origin of modern human behavior. *Journal of Human Evolution* 39(5):453–563.

McDougall, I., F. H. Brown, and J. G. Fleagle. 2005. Stratigraphic placement and age of modern humans from Kibish, Ethiopia. *Nature* 433:733–736.

McManus, D. A., and J. S. Creager. 1984. Sea-level data for the Bering-Chukchi shelves of Beringia from 19,000–10,000 14C yr B.P. *Quaternary Research* 21:317–325.

Medvedev, G. I. 1964. The place of the culture of Verkholenskaia Gora in the archeological sequence of the Baikal region. *American Antiquity* 29(4):461–466.

Meignen, L., and O. Bar-Yosef. 2002. The lithic industries of the Middle and Upper Paleolithic of the Levant: Continuity or break? *Archaeology, Ethnology and Anthropology of Eurasia* 3(11):12–21.

Meldgaard, J. 1962. *On the formative period of the Dorset culture.* Technical Paper No. 11. Calgary: Arctic Institute of North America.

Mellars, P. 1996. *The Neanderthal legacy: An archaeological perspective from western Europe.* Princeton, N.J.: Princeton University Press.

Meltzer, D. J. 1997. Monte Verde and the Pleistocene peopling of the Americas. *Science* 276:754–755.

———. 2001. Late Pleistocene cultural and technological diversity of Beringia: A view from down under. *Arctic Anthropology* 38(2):206–213.

Miller, D., and D. R. Bjornson. 1962. An investigation of cold injured soldiers in Alaska. *Military Medicine* 127:247–252.

Miller, N. F. 1984. The use of dung as fuel: An ethnographic example and an archaeological application. *Paléorient* 10:71–79.

Mochanov, Yu. A. 1969. Dyuktaiskaya verkhnepaleoliticheskaya kul'tura i nekotorye aspekty ee genezisa. *Sovetskaya arkheologiya* 4:235–239.

———. 1977. *Drevneishie etapy zaseleniya chelovekom Severo-Vostochnoi Azii.* Novosibirsk: Nauka.

———. 1983. *Arkheologicheskie pamyatniki Yakutii: Basseiny Aldana i Olkemy.* Novosibirsk: Nauka.

———. 1988. Drevneishii paleolit Diringa i problema vnetropicheskoi prarodiny chelovechestva. In A. N. Alekseev, L. T. Ivanova, and N. N. Kochmar, eds., *Arkheologiya Yakutii,* 15–54. Yakutsk: Yakutsk State University.

Mochanov, Yu. A., and S. A. Fedoseeva. 1996a. Dyuktai Cave. In F. H. West, ed., *American beginnings: The prehistory and palaeoecology of Beringia,* 164–174. Chicago: University of Chicago Press.

———. 1996b. Kukhtuy 3. In F. H. West, ed., *American beginnings: The prehistory and palaeoecology of Beringia,* 224–227. Chicago: University of Chicago Press.

———. 1996c. Berelekh, Allakhovsk Region. In F. H. West, ed., *American beginnings: The prehistory and palaeoecology of Beringia,* 218–222. Chicago: University of Chicago Press.

Mochanov, Yu. A., S. A. Fedoseeva, I. V. Konstantinov, N. V. Antipina, and V. G. Argunov. 1991. *Arkheologicheskie pamyatniki Yakutii: Basseiny Vilyuya, Anabara i Oleneka.* Moscow: Nauka.

Monahan, C. M. 1996. New zooarchaeological data from Bed II, Olduvai Gorge, Tanzania: Implications for hominid behavior in the early Pleistocene. *Journal of Human Evolution* 31:93–128.

Morlan, R. E. 1970. Wedge-shaped core technology in northern North America. *Arctic Anthropology* 7(2):17–37.

———. 1978a. Technological characteristics of some wedge shaped cores in northwestern North America and Northeast Asia. *Asian Perspectives* 19(1):96–106.

———. 1978b. Early Man in Northern Yukon Territory: Perspectives as of 1977. In A. L. Bryan, ed., *Early Man in America from a circum-Pacific perspective*, 78–95. Occasional Paper No. 1. Alberta: Department of Anthropology, University of Alberta.

———. 1986. Pleistocene archaeology in Old Crow Basin: A critical reappraisal. In A. L. Bryan, ed., *New evidence for the Pleistocene peopling of the Americas*, 27–48. Orono, Maine: Center for the Study of Early Man.

———. 1987. The Pleistocene archaeology of Beringia. In M. H. Nitecki and D. V. Nitecki, eds., *The evolution of human hunting*, 267–307. New York: Plenum.

Morlan, R. E., and J. Cinq-Mars. 1982. Ancient Beringians: Human occupation in the late Pleistocene of Alaska and the Yukon Territory. In D. M. Hopkins, J. V. Matthews Jr., C. E. Schweger, and S. B. Young, eds., *Paleoecology of Beringia*, 353–381. New York: Academic Press.

National Snow and Ice Data Center. Updated August 2006. Available at http://www.nsidc.org/sotc/permafrost.html (accessed September 2006).

Nelson, D. E., R. E. Morlan, J. S. Vogel, J. R. Southon, and C. R. Harington. 1986. New dates on northern Yukon artifacts: Holocene not Pleistocene. *Science* 232:749–751.

Nelson, N. 1935. Early migration of man to America. *Natural History* 35(4):356.

———. 1937. Notes on cultural relations between Asia and America. *American Antiquity* 2:264–272.

Nelson, R. E., and L. D. Carter. 1987. Paleoenvironmental analysis of insects and extralimital *Populus* from an early Holocene site on the Arctic Slope of Alaska. *Arctic and Alpine Research* 19:230–241.

Nelson, R. K. 1973. *Hunters of the northern forest*. Chicago: University of Chicago Press.

Nettle, D. 1999. Linguistic diversity of the Americas can be reconciled with a recent colonization. *Proceedings of the National Academy of Sciences* 96:3325–3329.

New Oxford Atlas. 1978. Rev. ed. Oxford: Oxford University Press.

Nichols, J. 1990. Linguistic diversity and the first settlement of the New World. *Language* 66:475–521.

Odess, D., and J. Rasic. 2004. Nogahabara 1 and the late Pleistocene prehistory of Alaska. Paper presented at the 32nd annual meeting of the Alaska Anthropological Association, Anchorage, Alaska, 10–12 March.

Orr, K. D., and D. C. Fainer. 1952. Cold injuries in Korea during winter of 1950–51. *Military Medicine* 31:177–220.

Osgood, C. 1940. *Ingalik material culture.* Publications in Anthropology No. 22. New Haven, Conn.: Yale University.

Oswalt, W. H. 1976. *The anthropological analysis of food-getting technology.* New York: Wiley.

———. 1987. Technological complexity: The Polar Eskimos and the Tareumiut. *Arctic Anthropology* 24(2):82–98.

Pankratova, I. 2006. The Ushki-5 site: New results of research, problems and prospects. Paper presented at "Pleistocene Human Colonization of Arctic and Subarctic Siberia and Beringia." Center for the Study of the First Americans, Texas A & M University, College Station, Texas, 17–19 November 2006.

Parfitt, S. A., and M. B. Roberts. 1999. Human modification of faunal remains. In M. B. Roberts and S. A. Parfitt, eds., *Boxgrove: A middle Pleistocene hominid site at Eartham Quarry, Boxgrove, West Sussex,* 395–415. Archaeological Report No. 17. London: English Heritage.

Pavlov, P., J. I. Svendsen, and S. Indrelid. 2001. Human presence in the European Arctic nearly 40,000 years ago. *Nature* 413:64–67.

Pavlova, E., V. V. Pitul'ko, M. A. Anisimov, A. E. Basilyan, and P. A. Nikolsky. 2006. Paleolithic sites of Northeast Asia: Their geomorphology and environment. Paper presented at "Pleistocene Human Colonization of Arctic and Subarctic Siberia and Beringia." Center for the Study of the First Americans, Texas A & M University, College Station, Texas, 17–19 November 2006.

Pearson, G. A. 1999. Early occupations and cultural sequence at Moose Creek: A late Pleistocene site in central Alaska. *Arctic* 52(4):332–345.

Pearson, G. A., and W. R. Powers. 2001. The Campus site re-excavation: New efforts to unravel its ancient and recent past. *Arctic Anthropology* 38(1): 100–119.

Peeters, L., and Ø. Totland. 1999. Wind to insect pollination ratios and floral traits in five alpine *Salix* species. *Canadian Journal of Botany* 77:556–563.

Peteet, D. M., and D. H. Mann. 1994. Late-glacial vegetational, tephra, and climatic history of southwestern Kodiak Island, Alaska. *Ecoscience* 1:255–267.

Petrin, V. T. 1986. *Paleoliticheskie pamyatniki Zapadno-Sibirskoi ravniny.* Novosibirsk: Nauka.

Péwé, T. L. 1975. *Quaternary Geology of Alaska.* Professional Paper No. 835. Washington, D.C.: U.S. Geological Survey.

Phillips, R. L., and E. M. Brouwers. 1990. Vibracore stratigraphy of the northeastern Chukchi Sea, Alaska. Paper presented at the 19th annual Arctic Workshop, Boulder, Colorado.

Phippen, P. G. 1988. Archaeology of Owl Ridge: A Pleistocene-Holocene boundary age site in central Alaska. M.A. thesis, University of Alaska, Fairbanks.

Pidoplichko, I. G. 1969. *Pozdnepaleoliticheskie zhilishcha iz kostei mamonta na Ukraine.* Kiev: Naukova Dumka.

———. 1976. *Mezhirichskie zhilishcha iz kostei mamonta.* Kiev: Naukova Dumka.

Pisaric, M. F. J., G. M. MacDonald, A. A. Velichko, and L. C. Cwynar. 2001. The late-glacial and post-glacial vegetation history of the northwestern limits of Beringia based on pollen, stomate and tree stump evidence. *Quaternary Science Reviews* 20:235–245.

Pitul'ko, V. V. 1993. An early Holocene site in the Siberian High Arctic. *Arctic Anthropology* 30(1):13–21.

———. 2001. Terminal Pleistocene–early Holocene occupation in Northeast Asia and the Zhokhov assemblage. *Quaternary Science Reviews* 20:267–275.

Pitul'ko, V. V., and A. K. Kasparov. 1996. Ancient arctic hunters: Material culture and survival strategy. *Arctic Anthropology* 33(1):1–36.

Pitul'ko, V. V., P. A. Nikolsky, E. Y. Girya, A. E. Basilyan, V. E. Tumskoy, S. A. Koulakov, S. N. Astakhov, E. Y. Pavlova, and M. A. Anisimov. 2004. The Yana RHS site: Humans in the Arctic before the Last Glacial Maximum. *Science* 303:52–56.

Potter, B. A. 2002. A provisional correlation of stratigraphy, radiometric dates, and archaeological components at the Gerstle River site, central Alaska. *Anthropological Papers of the University of Alaska, NS* 2(1):73–93.

Powers, W. R. 1973. Paleolithic Man in Northeast Asia. *Arctic Anthropology* 10(2):1–106.

———. 1983. Lithic technology of the Dry Creek site. In W. R. Powers, R. D. Guthrie, and J. F. Hoffecker, eds., *Dry Creek: The archeology and paleoecology of a late Pleistocene Alaskan hunting camp*, 62–181. Fairbanks: U.S. National Park Service.

Powers, W. R., R. D. Guthrie, and J. F. Hoffecker, eds. 1983. *Dry Creek: The archeology and paleoecology of a late Pleistocene Alaskan hunting camp.* Fairbanks: U.S. National Park Service.

Powers, W. R., and T. D. Hamilton. 1978. Dry Creek: A late Pleistocene human occupation in central Alaska. In A. L. Bryan, ed., *Early Man in America from a circum-Pacific perspective*, 72–77. Occasional Paper No. 1. Alberta: Department of Anthropology, University of Alberta.

Powers, W. R., and J. F. Hoffecker. 1989. Late Pleistocene settlement in the Nenana Valley, central Alaska. *American Antiquity* 54(2):263–287.

Qian, H., P. V. Krestov, P. Fu, Q. Wang, J.-S. Song, and C. Chourmouzis. 2003. Phytogeography of the Far East Asia. In J. Kolbek, M. Srutek, and E. Box, eds., *Forest vegetation of Northeast Asia*, 51–91. Dordrecht: Kluwer Academic.

Rainey, F. 1939. Archaeology in central Alaska. *Anthropological Papers of the American Museum of Natural History* 36(4):355–405.

———. 1940. Archaeological investigation in central Alaska. *American Antiquity* 4:299–308.

Rasic, J. 2003. Ancient hunters of the western Brooks Range: Integrating research and cultural resource management. *Alaska Park Science* winter:20–25.

Rasic, J., and R. Gal. 2000. An early lithic assemblage from the Tuluaq site, northwest Alaska. *Current Research in the Pleistocene* 17:66–68.

Reanier, R. E. 1995. The antiquity of Paleoindian materials in northern Alaska. *Arctic Anthropology* 32(1):31–50.

———. 1996. Putu and Bedwell. In F. H. West, ed., *American beginnings: The prehistory and paleoecology of Beringia*, 505–511. Chicago: University of Chicago Press.

Reference Book on Climate of the USSR, 1963–1969. 1970. Vol. 24, parts 2 and 4; Vol. 33, parts 2 and 4. Moscow: Gosmeteoizat.

Rhode, D., D. B. Madsen, and P. J. Brantingham. 2003. Preliminary assessment of the utility of dung as fuel during the initial occupation of Beringia, based on analogues from the Tibetan Plateau. Paper presented at the 16th Congress of the International Quaternary Association (INQUA), Reno, Nevada, July.

Richards, M. P., P. B. Pettitt, M. C. Stiner, and E. Trinkaus. 2001. Stable isotope evidence for increasing dietary breadth in the European Mid-Upper Paleolithic. *Proceedings of the National Academy of Sciences* 98:6528–6532.

Rick, J. 1980. *Prehistoric hunters of the high Andes.* New York: Academic Press.

Rightmire, G. P. 1998. Human evolution in the middle Pleistocene: The role of *Homo heidelbergensis. Evolutionary Anthropology* 6(6):218–227.

Ritchie, J. C., and L. C. Cwynar. 1982. The late Quaternary vegetation of the North Yukon. In D. M. Hopkins, J. V. Matthews Jr., C. E. Schweger, and S. B. Young, eds., *Paleoecology of Beringia*, 113–126. New York: Academic Press.

Roebroeks, W., and T. van Kolfschoten. 1995. The earliest occupation of Europe: A reappraisal of artefactual and chronological evidence. In W. Roebroeks and T. van Kolfschoten, eds., *The earliest occupation of Europe*, 297–315. Leiden: University of Leiden.

Rogachev, A. N., and A. A. Sinitsyn. 1982. Kostenki 15 (Gorodtsovskaya stoyanka). In N. D. Praslov and A. N. Rogachev, eds., *Paleolit Kostenkovsko-Borshchevskogo raiona na Donu 1879–1979*, 162–171. Leningrad: Nauka.

Rogers, R. A., L. D. Martin, and T. D. Nicklas. 1990. Ice-Age geography and the distribution of native North American languages. *Journal of Biogeography* 17:131–143.

Sablin, M. V., and G. A. Khlopachev. 2002. The earliest Ice Age dogs: Evidence from Eliseevichi I. *Current Anthropology* 43(5):795–799.

Sahnouni, M., and J. de Heinzelin. 1998. The site of Aïn Hanech revisited: New investigations at this Lower Pleistocene site in northern Algeria. *Journal of Archaeological Science* 25:1083–1101.

Savolainen, P., Y. Zhang, J. Luo, J. Lundeberg, and T. Leitner. 2002. Genetic evidence for an East Asian origin for domesticated dogs. *Science* 298:1610–1613.

Schaaf, J., ed. 1988. *The Bering Land Bridge National Preserve: An archaeological survey.* Anchorage: U.S. National Park Service.

Schurr, T. G., and S. T. Sherry. 2004. Mitochondrial DNA and Y chromosome diversity and the peopling of the Americas: Evolutionary and demographic evidence. *American Journal of Human Biology* 16:420–439.

Schweger, C. E. 1982. Late Pleistocene vegetation of eastern Beringia: Pollen analysis of dated alluvium. In D. M. Hopkins, J. V. Matthews Jr., C. E. Schweger, and S. B. Young, eds., *Paleoecology of Beringia,* 95–112. New York: Academic Press.

———. 1997. Late Quaternary Palaeoecology of the Yukon: A review. In H. V. Danks and J. A. Downes, eds., *Insects of the Yukon,* 59–72. Monograph Series No. 2. Ottawa: Biological Survey of Canada.

Scudder, G. G. E. 1997. Environment of the Yukon. In H. V. Danks and J. A. Downes, eds., *Insects of the Yukon,* 13–57. Monograph Series No. 2. Ottawa: Biological Survey of Canada.

Severinghaus, J. P., and E. J. Brook. 1999. Abrupt climate change at the end of the last glacial period inferred from trapped air in polar ice. *Science* 286: 930–934.

Shackleton, N. J., and N. D. Opdyke. 1973. Oxygen isotope and paleomagnetic stratigraphy of equatorial Pacific core V28–238: Temperatures and ice volumes on a 10^3 and 10^6 year scale. *Quaternary Research* 3:39–55

Shaver, G. R., and F. S. Chapin III. 1991. Production/biomass relationships and element cycling in contrasting arctic vegetation types. *Ecological Monographs* 61:1–31.

Shea, J. J. 1988. Spear points from the Middle Paleolithic of the Levant. *Journal of Field Archaeology* 15:441–450.

Shen, G., W. Wang, Q. Wang, J. Zhao, K. Collerson, C. Zhou, and P. V. Tobias. 2002. U-series dating of Liujiang hominid site in Guangxi, southern China. *Journal of Human Evolution* 43:817–829.

Sher, A., A. Alfimov, D. Berman, and S. Kuzmina. 2002. Attempted reconstruction of seasonal temperature for the late Pleistocene arctic lowlands of northeastern Siberia based on fossil insects. Sixth Quaternary Environment of the Eurasian North (QUEEN) Workshop, Programme and Abstracts, 50.

Shilo, N. A., N. N. Dikov, and A. V. Lozhkin. 1967. Pervye dannye po stratigrafii paleolita Kamchatki. In *Istoriya i kul'tura narodov Severa Dal'nego Vostoka,* 32–41. Moscow: Nauka.

Sinitsyn, A. A. 2002. Nizhnie kul'turnye sloi Kostenok 14 (Markina gora) (Raskopki 1998–2001 gg.). In A. A. Sinitsyn, V. Ya. Sergin, and J. F. Hoffecker, eds., *Kostenki v kontekste paleolita Evrazii,* 219–236. St. Petersburg: Russian Academy of Sciences.

Skarland, I., and C. J. Keim. 1958. Archaeological discoveries on the Denali Highway. *Anthropological Papers of the University of Alaska* 6(2):79–88.

Slobodin, S. 1999. Northeast Asia in the late Pleistocene and early Holocene. *World Archaeology* 30(3):484–502.

———. 2001. Western Beringia at the end of the Ice Age. *Arctic Anthropology* 38(2):31–47.

Slobodin, S. B., and M. L. King. 1996. Uptar and Kheta: Upper Paleolithic sites of the Upper Kolyma region. In F. H. West, ed., *American beginnings: The prehistory and paleoecology of Beringia*, 236–244. Chicago: University of Chicago Press.

Smith, P. S. 1937. Certain relations between northwestern America and northeastern Asia. In G. G. McCurdy, ed., *Early Man*, 85–92. Philadelphia: Lippincott.

Smith, T. A. 1985. Spatial analysis of the Dry Creek site. *National Geographic Research Reports* 19:6–11.

Soffer, O. 1985. *The Upper Paleolithic of the Central Russian Plain*. San Diego: Academic Press.

———. 2000. Gravettian technologies in social contexts. In W. Roebroeks, M. Mussi, J. Svoboda, and K. Fennema, eds., *Hunters of the Golden Age: The Mid Upper Paleolithic of Eurasia 30,000–20,000 BP*, 59–75. Leiden: University of Leiden.

Solecki, R. S. . 1996. Prismatic core sites on the Kukpowruk and Kugurok Rivers. In F. H. West, ed., *American beginnings: The prehistory and paleoecology of Beringia*, 513–521. Chicago: University of Chicago Press.

Stanek, W., K. Alexander, and C. S. Simmons. 1981. *Reconnaissance of vegetation and soils along the Dempster Highway, Yukon Territory. I: Vegetation Types*. Information Report No. BC-X-293. Victoria, B.C.: Canadian Forestry Service, Pacific Forest Research Centre.

Stanley-Brown, J. 1892. Geology of the Pribilof. *Geological Society of America Bulletin* 3:496–500.

Steffian, A. F., E. P. Eufemio, and P. G. Saltonstall. 2002. Early sites and microblade technologies from the Kodiak Archipelago. *Anthropological Papers of the University of Alaska, NS* 2(1):1–35.

Stepanova, M. V., I. S. Gurvich, and V. V. Khramova. 1964. The Yukagirs. In M. G. Levin and L. P. Potapov, eds., *The peoples of Siberia*, 788–798. Chicago: University of Chicago Press.

Stephenson, R. O., S. C. Gerlach, R. D. Guthrie, C. R. Harington, R. O. Mills, and G. Hare. 2001. Wood bison in late Holocene Alaska and adjacent Canada: Paleontological, archaeological and historical records. In S. C. Gerlach and M. S. Murray, eds., *People and wildlife in northern North America: Essays in honor of R. Dale Guthrie*, 125–159. International Series No. 994. Oxford: British Archaeological Reports.

Stiner, M. C., N. D. Munro, T. A. Surovell, E. Tchernov, and O. Bar-Yosef. 1999. Paleolithic population growth pulses evidenced by small animal exploitation. *Science* 283:190–194.

Straus, L. G. 2000. Solutrean settlement of North America? A review of reality. *American Antiquity* 65(2):219–226.

Stringer, C., and C. Gamble. 1993. *In search of the Neanderthals: Solving the puzzle of human origins.* New York: Thames and Hudson.

Stringer, C., and R. McKie. 1996. *African exodus: The origins of modern humanity.* New York: Holt.

Stringer, C., and E. Trinkaus. 1999. The human tibia from Boxgrove. In M. B. Roberts and S. A. Parfitt, eds., *Boxgrove: A Middle Pleistocene hominid site at Eartham Quarry, Boxgrove, West Sussex,* 420–422. London: English Heritage.

Sumner, D. S., T. L. Criblez, and W. H. Doolittle. 1974. Host factors in human frostbite. *Military Medicine* 139:454–461.

Svoboda, J., V. Lozek, and E. Vlcek. 1996. *Hunters between East and West: The Paleolithic of Moravia.* New York: Plenum.

Szeicz, J. M., and G. M. MacDonald. 2001. Montane climate and vegetation dynamics in easternmost Beringia during the late Quaternary. *Quaternary Science Reviews* 20:247–257.

Tabarev, A. V. 1997. Paleolithic wedge-shaped microcores and experiments with pocket devices. *Lithic Technology* 22(2):139–149.

Ten Brink, N., and C. F. Waythomas. 1985. Late Wisconsin glacial chronology of the north-central Alaska Range: A regional synthesis and its implications for early human settlements. *National Geographic Society Research Reports* 19:15–32.

Théry-Parisot, I. 2001. *Economie des combustibles au Paléolithique.* Paris: CNRS Editions.

Thieme, H. 1997. Lower Palaeolithic hunting spears from Germany. *Nature* 385:807–810.

Thompson, R. M. 1948. Notes on the archaeology of the Utukok River. *American Antiquity* 14(1):62–65.

Thorson, R. M. 1983. Stratigraphic reconnaissance of the Chugwater Site, Alaska. Report on file, U.S. Army Corps of Engineers, Alaska District, Anchorage.

Thorson, R. M., and G. Bender. 1985. Eolian deflation by ancient katabatic winds: A late Quaternary example from the north Alaska Range. *Geological Society of America Bulletin* 96:702–709.

Thorson, R. M., and T. D. Hamilton. 1977. Geology of the Dry Creek site: a stratified Early Man site in interior Alaska. *Quaternary Research* 7:149–176.

Titlyanova, A. A., V. I. Kiryushin, and I. P. Okhin'ko. 1984. *Agrotsenozy stepnoi zony.* Novosibirsk: Nauka.

Ton-That, T., B. Singer, and M. Paterne. 2001. $^{40}Ar/^{39}Ar$ dating of latest Pleistocene (41 ka) marine tephra in the Mediterranean Sea: Implications for global climate records. *Earth and Planetary Science Letters* 184(3–4):645–658.

Trinkaus, E. 1981. Neanderthal limb proportions and cold adaptation. In C. Stringer, ed., *Aspects of human evolution,* 187–224. London: Taylor and Francis.

Troeng, John. 1993. Worldwide chronology of fifty-three prehistoric innovations. *Acta Archaeologica Lundensia* 8(21).

Tseitlin, S. M. 1979. *Geologiya paleolita Severnoi Azii*. Moscow: Nauka.

Turner, C. G. 1990. Paleolithic teeth of the central Siberian Altai Mountains. In *Chronostratigraphy of the Paleolithic in North, Central, East Asia and America*, 239–243. Novosibirsk: USSR Academy of Sciences.

———. 2002. Teeth, needles, dogs, and Siberia: Bioarchaeological evidence for the colonization of the New World. In N. G. Jablonski, ed., *The first Americans*, 123–158. Memoirs No. 27. San Francisco: California Academy of Sciences.

Turner, E. 1999. The problems of interpreting hominid subsistence strategies at Lower Palaeolithic sites: A case study from the central Rhineland of Germany. In H. Ullrich, ed., *Hominid evolution: Lifestyles and survival strategies*, 365–382. Gelsenkirchen-Schwelm: Edition Archaea.

Tuttle, R, H. 1986. *Apes of the world: Their social behavior, communication, mentality, and ecology*. Park Ridge, N.J.: Noyes.

Vasil'ev, S. A. 1996. *Pozdnii paleolit verkhnego Eniseya*. St. Petersburg: Russian Academy of Sciences.

———. 2001. The final Paleolithic in northern Asia: Lithic assemblage diversity and explanatory models. *Arctic Anthropology* 38(2):3–30.

Vasil'ev, S. A., O. V. Kuznetsov, and M. N. Meshcherin. 1987. Poselenie Tolbaga (novyi etap issledovanii). In *Prirodnaya i drevnii chelovek v pozdnem antropogene*, 109–121. Ulan-Ude: Nauka.

Vekua, A., et al. 2002. A new skull of early *Homo* from Dmanisi, Georgia. *Science* 297:85–89.

Velichko, A. A. 1984. Late Pleistocene spatial paleoclimatic reconstructions. In A. A. Velichko, ed., *Late Quaternary environments of the Soviet Union*, 261–285. Minneapolis: University of Minnesota Press.

Vereshchagin, N. K. 1977. Berelekhskoe "kladbishche" mamontov. *Trudy Zoologicheskogo instituta AN SSSR* 72:5–50.

———. 1979. Ostatki mlekopitayushchikh iz paleoliticheskogo sloya VI stoyanki Ushki 1. In *Novye arkheologicheskie pamyatniki Severa Dal'nego Vostoka (po dannym Severo-Vostochno-Aziatskoi Kompleksnoi Arkheologicheskoi Ekspeditsii)*, 18–19. Magadan: USSR Academy of Sciences.

Vereshchagin, N. K., and G. F. Baryshnikov. 1982. Paleoecology of the mammoth fauna in the Eurasian Arctic. In D. M. Hopkins, J. V. Matthews Jr., C. E. Schweger, and S. B. Young, eds., *Paleoecology of Beringia*, 267–279. New York: Academic Press.

———. 1984. Quaternary mammalian extinctions in northern Eurasia. In P. S. Martin and R. G. Klein, eds., *Quaternary extinctions*, 483–516. Tucson: University of Arizona Press.

Vereshchagin, N. K., and I. E. Kuz'mina. 1984. Late Pleistocene mammal fauna of Siberia. In H. E. Wright and C. Barnowsky, eds., *Late Quaternary environments of the Soviet Union*, 219–222. Minneapolis: University of Minnesota Press.

Vereshchagin, N. K., and Yu. A. Mochanov. 1972. Samye severnye v mire sledy verkhnego paleolita (Berelekhskoe mestonakhozhdenie v nizov'akh r. Indigirki). *Sovetskaya arkheologiya* 3:332–336.

Vereshchagin, N. K., and V. V. Ukraintseva. 1985. Proiskhozhdenie i stratigrafiya Berelekhskogo "kladbishcha" mamontov. *Trudy Zoologicheskogo instituta AN SSSR* 131:104–113.

Viereck, L. A., and E. L. Little Jr. 2007. *Alaska trees and shrubs*. Fairbanks: University of Alaska Press.

Villa, P., F. Bon, and J.-C. Castel. 2002. Fuel, fire and fireplaces in the Palaeolithic of western Europe. *Review of Archaeology* 23(1):33–42.

Vinson, D. M. 1993. Taphonomic analysis of faunal remains from Trail Creek Caves, Seward Peninsula, Alaska. M.A. thesis, Department of Anthropology, University of Alaska, Fairbanks.

Wahl, H. E., D. B. Fraser, R. C. Harvey, and J. B. Maxwell. 1987. *Climate of the Yukon*. Climatological Studies No. 40. Ottawa: Atmospheric Environment Service, Environment Canada.

Wahrhaftig, C. 1958. *Quaternary geology of the Nenana River Valley and adjacent parts of the Alaska Range*. Professional Paper No. 293. Washington, D.C.: U.S. Geological Survey.

——. 1970. *Geologic map of the Healy D-5 Quadrangle*. Washington, D.C.: U.S. Geological Survey.

Walker, A., and P. Shipman. 1996. *The wisdom of the bones: In search of human origins*. New York: Knopf.

Walker, D. A., J. G. Bockheim, F. S. Chapin III, W. Eugster, F. E. Nelson, and C. L. Ping. 2001. Calcium-rich tundra, wildlife, and the "Mammoth Steppe." *Quaternary Science Reviews* 20:149–163.

Wallace, A. 1876. *The geographical distribution of animals*. 2 vols. New York: Harper.

Ward, B. C., M. C. Wilson, D. W. Nagorsen, D. E. Nelson, J. C. Driver, and R. J. Wigen. 2003. Port Eliza Cave: North American west coast interstadial environment and implications for human migrations. *Quaternary Science Reviews* 22:1383–1388.

Waters, M., R., S. L. Forman, and J. M. Pierson. 1997. Diring Yuriakh: A Lower Paleolithic site in central Siberia. *Science* 275:1281–1284.

West, F. H. 1967. The Donnelly Ridge site and the definition of an early core and blade complex in central Alaska. *American Antiquity* 32:360–382.

——. 1975. Dating the Denali complex. *Arctic Anthropology* 12(1):76–81.

——. 1981. *The archaeology of Beringia*. New York: Columbia University Press.

———. 1996a. The archaeological evidence. In F. H. West, ed., *American beginnings: The prehistory and paleoecology of Beringia*, 537–559. Chicago: University of Chicago Press.

———. 1996b. South Central Alaska Range: Tangle Lakes region: Introduction. In F. H. West, ed., *American beginnings: The prehistory and paleoecology of Beringia*, 375–380. Chicago: University of Chicago Press.

West, F. H., B. S. Robinson, and M. L. Curran. 1996. Phipps site. In F. H. West, ed., *American beginnings: The prehistory and paleoecology of Beringia*, 381–386. Chicago: University of Chicago Press.

West, F. H., B. S. Robinson, and C. F. West. 1996. Whitmore Ridge. In F. H. West, ed., *American beginnings: The prehistory and paleoecology of Beringia*, 386–394. Chicago: University of Chicago Press.

Whittaker, R. H. 1975. *Communities and ecosystems.* 2nd ed. New York: Macmillan.

Wiken, E. B., D. M. Welch, G. R. Ironside, and D. G. Taylor. 1981. *The Northern Yukon: An ecological land survey.* Ecological Land Classification Series No. 6. Ottawa: Environment Canada.

Willey, G. R. 1966. *An introduction to American archaeology.* Vol. 1: *North and Middle America.* Englewood Cliffs, N.J.: Prentice Hall.

Wilmsen, E. N. 1964. Flake tools in the American Arctic: Some speculations. *American Antiquity* 29(3):338–344.

———. 1965. An outline of Early Man studies in the United States. *American Antiquity* 31:172–192.

Wilson, M. C., and J. A. Burns. 1999. Searching for the earliest Canadians: Wide corridors, narrow doorways, small windows. In R. Bonnichsen and K. L. Turnmire, eds., *Ice Age peoples of North America: Environments, origins, and adaptations of the first Americans*, 213–248. Corvallis: Oregon State University Press.

Workman, W. 2001. Reflections on the utility of the coastal migration hypothesis in understanding the peopling of the New World. Paper presented at the Alaska Anthropological Association annual meeting, Fairbanks, Alaska, March 21–24.

Wormington, H. M. 1957. *Ancient man in North America.* 4th ed. Popular Series No. 4. Denver, Colo.: Denver Museum of Natural History.

Wright, M. 1986. Le Bois de Vache: This chip's for you. *Alberta Archaeological Review* 12:3–6.

Yang, D., and T. Ohata. 2001. A bias corrected Siberian regional precipitation climatology. *Journal of Hydrometeorology* 2:122–139.

Yesner, D. R. 1995. Human adaptation at the Pleistocene-Holocene boundary (circa 13,000 to 8,000 BP) in eastern Beringia. In L. G. Straus, B. V. Eriksen, J. M. Erlandson, and D. R. Yesner, eds., *Humans at the end of the Ice Age*, 255–276. New York: Plenum.

———. 2001. Human dispersal into interior Alaska: Antecedent conditions, mode of colonization, and adaptations. *Quaternary Science Reviews* 20:315–327.

Yesner, D. R., C. M. Barton, G. A. Clark, and G. A. Pearson. 2004. Peopling of the Americas and continental colonization: A millennial perspective. In C. M. Barton, G. A. Clark, D. R. Yesner, and G. A. Pearson, eds., *The settlement of the American continents: A multidisciplianry approach to human biogeography*, 196–213. Tucson: University of Arizona Press.

Yi, S., and G. Clark. 1985. The "Dyuktai Culture" and New World origins. *Current Anthropology* 26(1):1–20.

Young, S. B. 1994. *To the Arctic: An introduction to the Far Northern World*. New York: Wiley.

Yurtsev, B. A. 1982. Relics of the xerophyte vegetation of Beringia in northeastern Asia. In D. M. Hopkins, J. V. Matthews Jr., C. E. Schweger, and S. B. Young, eds., *Paleoecology of Beringia*, 157–177. New York: Academic Press.

———. 1984. Problems of the late Cenozoic paleogeography of Beringia in light of phytogeographic evidence. In V. L. Kontrimavichus, ed., *Beringia in the Cenozoic era*, 129–153. New Delhi: Amerind.

———. 2001. The Pleistocene "tundra-steppe" and the productivity paradox: The landscape approach. *Quaternary Science Reviews* 20:165–174.

Zhu, R. X., K. A. Hoffman, R. Potts, C. L. Deng, Y. X. Pan, B. Guo, C. D. Shi, Z. T. Guo, B. Y. Yuan, Y. M. Hou, and W. W. Huang. 2001. Earliest presence of humans in Northeast Asia. *Nature* 413:413–417.

Zimov, S. A., V. I. Chuprynin, A. P. Oreshko, F. S. Chapin III, J. F. Reynolds, and M. C. Chapin. 1995. Steppe-tundra transition: A herbivore-driven biome shift at the end of the Pleistocene. *American Naturalist* 146:765–794.

Index